Metal Detecting For Beginners

How To Get Started Correctly

by Ed Tisdale

AF709907

Published by:
Ed Tisdale
Cape Coral, Florida 33904

Original © copyright 2001
Revised Edition © copyright 2003
ISBN 0-9744166-0-6

by Edward W. Tisdale
(All Rights Reserved)

Dedication

To my patient, understanding, and wonderful wife Nancy who enjoys wearing some of the better pieces of jewelry I sometimes bring home, who puts up with the mess in the sink from initial cleaning of finds, and who understands my passion for treasure hunting.

To my friend Norm Jones, a Yorkshire man by birth. Norm initiated me into the hobby, convinced me to spend the money to buy a good detector, and arranged for my wife and I to visit he and wife Betty at their retirement home in Nostell Priory, Yorkshire, UK for my first "PRIMO" treasure hunting expedition out of the country. Norm begins every trip with, "Well, it's out there! All we have to do is go find it."

To Harold Kelly. Harold sold Norm his first detector and encouraged him to excel in the hobby. Harold is the only person I know who found a hoard of Roman silver and gold coins. You will read more about this later in the book. Harold passed away September, 2000. We will miss your bright smile and terrific sense of humor. Thank you for taking Norm and I to your favorite Yorkshire sites and cementing our friendships forever.

To my grandchildren, present and future. Hopefully you'll help me dig the targets when I get too old … if you aren't too busy recovering your own.

To all the new friends I've made through the hobby of metal detecting and metal detecting Internet forums. Keep filling in the holes and properly disposing of the trash found among the treasures.

To Bob Davis, Republic, Missouri, an avid detectorist for over 30 years. Bob spent hours in thought to provide much appreciated suggestions for improving readability of this book, editing assistance, and contributed to its content. He also gave me a new lucky detecting hat that I proudly wear. Thank you Bob!

Table Of Contents

About The Author

I became involved in the hobby of metal detecting in 1991. Enjoyment of the hobby began in 1992. I'll explain why later. You might say I've become an avid hobbyist even though by some standards, I'm still considered new to the hobby by those who've treasure hunted since the 60's, some with "kit built" metal detectors.

Ed Tisdale, Author

Norm Jones, a friend made through Rotary Club, fascinated me with Roman and Medieval coins, musket balls, and jewelry from the same periods he had found while detecting in England near where he was born.

Never before had I been privileged to hold something so old from history in my hand. This was "look but don't touch" stuff you see protected in museums behind glass. Norm literally spent hours patiently answering questions about the hobby, the equipment he uses, and the wide range of equipment available.

Like most people, money was an issue that did not allow me to go out and buy a top of the line metal detector for personal fun and pleasure. I had left the employment and security of a large corporation to go into business for myself. All available money had to go into growing the business. A specialty wholesaler of lots of different unrelated items mailed us a special that included an inexpensive "no name" metal detector. I think I paid $119 wholesale. This appeared to be an opportunity to justify adding a rental item to inventory that I could use on my day off. Wow, was that a mistake.

On the first trip, Norm was walking along behind to coach me on technique and recovered the items my detector wouldn't signal on at three inches. After a couple more trips using this cheap detector, my interest in the hobby waned. On each trip, Norm would let me use his White's XLT while he checked out the settings on my machine.

A year later, I finally made the plunge and bought a demonstrator XLT from a catalog / retail operation in another town. I became truly hooked on the hobby over the next few months and began putting the interesting items I had found on display in the store and some nice looking beach found jewelry on my wife's fingers.

I am constantly learning something new about the hobby from every treasure hunting trip, my detecting friends, metal detector manufacturers, books, magazines and the Internet.

My wife and I own a hardware store in Cape Coral, Florida. We sell metal detectors in the store and believe it to be as natural of a merchandise category for us as plumbing, electrical, tools, paint, and hurricane storm panels. All of these departments require someone who has complete product knowledge and the user experience necessary to teach others how to effectively use the products we sell. Several of our employees also have become good detectorists.

Metal detectors were added due to our display of coins, jewelry, relics and artifacts from my metal detecting trips here in the states as well as to the UK. Customers kept asking where they could buy a metal detector that would find such valuable and interesting items. We used to direct people to a catalog house and another store where metal detectors could be purchased.

One day I remembered we ran a retail store and ought to be selling metal detectors ourselves. Its amazing how long it can take a person with a college degree in business management and an ex- marketing manager of a $2.5 Billion company to realize he's missing a great selling opportunity and the chance to mix pleasure with business.

The sole purpose of this book is to help the beginning detectorist get started correctly and avoid making mistakes that cost time and money. Along the way, we will attempt to entertain you, humor you, share personal metal detecting experiences, and make you as strong of an advocate of our fascinating hobby as you can possibly be.

We want to help eliminate much of the learning curve and to make you a more productive skilled hobbyist in a shorter period of time. We will share the mistakes most all of us "experienced" treasure hunters have made or heard about over the years. The wide selection of metal detector and treasure hunting equipment can be confusing to someone new to the hobby. We want this book to help you make the best decisions as to what you really need to buy, and to help you learn to use your equipment effectively.

Rationale or Justification?

The decision to finally buy that metal detector you want can be difficult. The difficulty is compounded by variables worthy of consideration *far* beyond the wide selection of manufacturers and the different models offered. The most difficult considerations to overcome almost always involve the spouse (him or her)! The most notable considerations can include:

****On a scale of 1 to 10, with 10 being certain divorce, where am I currently on the scale?***
****Will I enjoy my new metal detector long enough to outlast the cold shoulder and silence?***
****How upset will my spouse become if I tell the whole truth about how much it cost?***

After due consideration, most of us will make the decision to take the plunge by 1.) purchasing our first detector, 2.) selling the old one and buying a more expensive new one, or 3.) buying an additional metal detector.

Number three is not a decision for the faint of heart! As a service to men and women everywhere who have made the decision to make a metal detector purchase, realizing their spouse may not be truly understanding of the need, the following tips are offered.

Tip 1. Tell the spouse your doctor wrote you a prescription for the purchase of a metal detector because being out-of-doors, walking slowly, doing deep knee bends, and digging small holes is good for you.

Tip 2. Perhaps the most popular tip, "Gee whiz Honey, ***it's not just for me***, I bought the detector so that we could spend more quality time together… you know, rediscovering our love for each other."

Tip 3. Pay cash. Leave no "paper" trail to discovery. Keep the new detector at your treasure hunting friends house. When its time to go detecting, tell the spouse you will be going on your weekly nature walk and immediately ask where the mosquito repellent is kept.

Tip 4. Tell the spouse, "That guy at the detector store certainly is not a rocket scientist and look at all the neat stuff he found. I bet you and I could easily pay for this new metal detector in no time at all."

Tip 5. Should you actually dare to bring your new detector home in the factory fresh carton, and are discovered before it can be hidden, you might try, "Gosh Honey, that's not a **new** detector. That's my old one. I've had it for years. I just got it back from reconditioning. HOLY COW! Would you just look at what those guys charged. What bandits they are! From now on I'll just buy a new one."

Tip 6. An alternative to tip 5 would be, "Why Honey, I can't believe you haven't noticed my metal detector in the closet before. I bought this soooo long ago before we were married and I figured its about time I got it out and learned how to use it. Isn't it cool looking?"

Tip 7. "Yes Dear, I certainly do realize your car has been broken down for over three months. With this new metal detector I'll be able to find all those expensive parts I lost when I was taking it apart to fix it. I just wish you were as good at saving us money as I am!"

Tip 8. "No Honey, its not mine. It belongs to (fill in your best friends name). You know his / her spouse isn't nearly as loving and understanding as you, so he / she has to hide the metal detector here. Now please don't say anything about it in front of them and perhaps cause them to get into a big argument." (See tip #3)

I remember when I brought home my first good detector and told Nancy how much "***we had invested in the XLT***." The word "***invest"*** had up to that point in time been synonymous with the words "buy" and "purchase" in Nancy's shopping vocabulary. I then made a promise to myself that I would find some better way to deal with telling the whole truth in the future.

Several years later, and after giving Nancy some nice jewelry recovered on the beach, I brought home a SurfMaster PI Diver to begin surf treasure hunting. She smiled and said, "Great, lets head to the beach. I'll walk, swim, sip my soda, and read a book while you are finding me a lot more nice jewelry."

Detecting has in fact helped Nancy and I spend more quality time together. She loves the beach while I had always avoided going until I started treasure hunting. I encourage you to get your spouse involved in the hobby by not "hogging" all the machine time until you can justify buying your second machine.

Beginners Luck

Regardless of the activity being fishing or metal detecting, the expression "*beginners luck*" often describes a better catch or more valuable recovery than the "experienced" person had on the same trip!

Perhaps the experienced fisherman helps by suggesting the right lure and spot in which to cast and the experienced detectorist suggests the "newbie" start their search in a particular area and helps them set up their metal detector "just right" for conditions.

From the stories we hear, the above is not the case when metal detecting. Here is an example of what we think happens more often.

One day we were ringing up a metal detector sale for a newbie when the customer next in line said, "Hope you have beginners luck like my nephews did!" Naturally the newbie asked to hear the man's beginners luck story.

He said, "My nephews, came to visit me this past spring. They saw my metal detector in the garage and kept bugging me to take them treasure hunting. All they could talk about until I could take them during the next weekend was the pirate treasure they were certain to find.

As we walked across the sand to be further away from the sun bathers, I was dropping change from my pocket to be sure the boys would find something and not be disappointed with the hobby. As expected, less than forty five minutes later, the boys were ready to go swimming and put away the detector until later.

I asked the 8 year old if he was able to find the pirate treasure. He said no but he did find some money and showed me about half of the coins I had scattered earlier and a few more that had been on the beach much longer. The 6 year old told me, "Look at what I found Uncle Jack" and reached into his pocket and pulled out just about all of the rest of the change I had dropped … along with a beautiful, sparkling, big chunky, expensive looking man's gold bracelet."

I asked my nephew if he found the bracelet for me since it was too big for him. He said, "Oh no Uncle Jack, I'm going to give it to Daddy, because you can find your own anytime you want."

The uncle finished his story saying, "I detected up north for over 10 years and here on our beaches for about 5 years and still haven't found anything on our beaches that valuable."

A few weeks after buying his metal detector, a customer came in to proudly show what he had found the first and second time he had ever gone detecting. One trip was to a local beach, the other to a beach on the east coast of Florida.

The customer had found a large gold Indian head shaped ring with a diamond, rubies, and emeralds set into the Indian headdress band. Also found was a man's wedding band with three small diamonds, two gold pendants both set with diamonds, rubies and / or emeralds, and a woman's ring with about a 1.5k pear shaped diamond surrounded by diamond baguettes. The large diamond had been appraised at a little over $10,000.

Beginners luck jewelry found on the first two trips!

I read about a man's son, home on leave from military service in the United Kingdom who joined his father and his friends detecting a field. With a borrowed detector the son set off to an area of the field where he wouldn't be in anyone's way as he learned how to find and recover targets. Yep, you guessed it! He found what ended up being a hoard of Roman coins worth several hundred thousand dollars.

Beginners luck happens more often than experienced treasure hunters like to admit. Sure it helps to be experienced and knowledgeable in the hobby. It's just as important that you be lucky!

How Metal Detectors Work

I'm not an electronic "techno-geek" so you electrical engineers out there, please pardon me for this simplistic explanation as to how metal detectors help us find treasure.

VLF (Very Low Frequency) metal detectors transmit and receive a signal from the detectors search coil (loop) as it sweeps over the ground. When a metal object is present under the loop, the transmitted signal measures the relative conductivity (electrical phase) of the metal and returns the signal back to the loops receiving coil.

To simplify the explanation of a metals relative conductivity or phase, imagine a straight line with –95 being assigned to the left end of the line, 0 in the middle, and +95 being assigned to the right hand end of the line.

-95 ———————————— 0 ————————————+95

Ferrous metals such as iron and steel will be picked up in the minus (-) side of the scale. Precious metals, beginning with tiny pieces of aluminum foil and including gold, silver, platinum, brass, bronze, etc. (anything a magnet won't attract) will read in the range of positive numbers (+).

The signal is processed to emit a tone from the built in speaker or headphones plugged into the detector. Mid to higher priced detectors often provide a signal tone and a visual LCD or analog reading of what type of metal you have targeted. Some detectors will also tell you how deep the target may be in the ground.

VLF detectors allow the operator to set the level of discrimination desired to help eliminate receiving signals from trash targets such as pieces of ferrous metals (nails, bobby pins, and bits of steel and galvanized wire). Anything with a similar conductivity, below the operators set level of discrimination line, will be ignored or give you a broken / scratchy sounding signal. Anything above the minimum discrimination setting will cause a modest to strong signal to alert the operator of a potential desired target under the loop. More on discrimination settings is provided later in this book.

Lower priced detectors may only provide a signal tone based on the discrimination setting. Mid priced detectors generally provide the tone and visual identification of the target assigned to several conductivity identification zones. A top of the line metal detector will provide one or more tones and assign some type of individual target VDI (Visual Display Indicator) reference number and may show Icons as to what the target may be.

As examples, the White's XLT Spectrum on a clad U.S. quarter will generally give you a VDI of +83 to +86 and a Icon of a quarter will appear on the LCD screen. A small aluminum pull tab from a can, a U.S. nickel, and a small gold ring all have very similar conductivity in the ground giving a VDI of +18 to +39. The Icons shown will be pull tab and ring, or ring and pull tab. The first Icon shown, if two are displayed, is the detectors "best guess" as to what the target may be.

All but the lowest priced detectors have a pinpoint switch which allows you to instantly shift the machine from discriminate to an "all metal" mode and reduce the size of the signal to pinpoint the item under the center of the loop. As mentioned earlier, a top of the line metal detector will tell you the depth of the target when in pinpoint mode.

Generally speaking, a detectors signal penetrates the ground a bit deeper than the loop is big. Older items may set up a halo or aura effect from natural galvanic corrosion allowing the detector to read the halo about the same as the metal that caused it. The result is that older undisturbed items can sometimes be found at greater than normal depths although with a somewhat weaker signal.

Multiple frequency metal detectors for land or water use have become popular to those interested in expanding their top of the line arsenal of metal detecting equipment.

As its description implies, multiple frequency metal detectors use more than one frequency when transmitting and receiving the signal through the search coil. On some machines, the user can choose to use one or more of the available frequencies to identify the target, allow the metal detector to choose the frequency to receive the greatest target reading depth, or receive a signal based on the best data from all frequencies being used. Multiple frequency detectors are known to provide a more stable signal when detecting highly mineralized ground.

Some multiple frequency metal detector brands are far more complex and difficult to learn how to use than others. When considering a multiple frequency metal detector, regardless of being an experienced or beginning treasure hunter, give consideration not only to cost, but also as to how complex (difficult) it may be to operate and interpret the information it provides. A complex multi frequency metal detector is generally not recommended for the beginning treasure hunter.

Pulse Induction (PI) metal detectors operate in a similar method as RADAR on an airplane or boat. The loop transmits a signal and receives a return signal looking for a change in density anywhere within the range of the signal. As the PI detectors loop passes over a metal item, a change in density is detected and an audible signal is heard in the headphones.

A RADAR processes the signal as a blip on a screen. A PI metal detector processes the signal as an audible tone. Pulse induction metal detectors usually do not allow the operator to discriminate out unwanted metal targets such as small iron and steel (nails, wire, bobby pins, etc.). Pulse induction detectors are used primarily for underwater use and can penetrate the ground or bottom surface deeper than most single frequency VLF detectors.

A PI metal detector's signal is enhanced by salt minerals within salt water detecting conditions providing even greater signal depth than in dry ground / no salt conditions. A PI detector purchased for use in deep water should have a weighted loop to make it easier to keep the loop placed firmly on the bottom while diving. It is hard enough to fight the water resistance against the down rod and loop when swinging the loop side to side in the water. You don't want to expend any extra energy also having to constantly press a buoyant non weighted loop to the bottom. A buoyant loop is usually not a big problem while wading up to chest deep water.

Loop (Coil) Size and Configuration

Some folks call them loops, others refer to them as coils. I most often call them loops because that's what "the round thing" looks like to me. Regardless of what you call them, the loop or coil is the part of your detector that you keep as near the surface of the ground as possible in hopes of generating a signal when it is swept over a desirable target.

Standard loop sizes on the majority of metal detectors will generally be about 8" to 10". Most detector manufacturers offer larger as well as smaller loops as accessory items for their mid range to top of the line machines. Why would you want or need a larger or smaller loop?

As we stated earlier, the metal detectors signal will generally penetrate the ground a little deeper than the loop is big. Most top of the line detectors allow an experienced user to "tweak" various elements of the signal to help it penetrate the ground even further than it can with the factory preset / suggested settings. This is particularly advantageous when dealing with highly mineralized soil conditions. Effects of ground mineralization will be discussed later.

It stands to reason that one would expect to receive signals from deeper items using a 15" loop than from a 10" loop. This is true, ***but not always the whole truth***.

A bigger loop of 15" or larger, will pick up *larger* items deeper in the ground than would a 10" loop. The surprise is that the 10" loop will generally do a better job receiving signals on *smaller* targets within its range than the 15" loop would do within the same range.

A larger loop increases the users difficulty in pinpointing and differentiating between multiple targets under the loop. A larger loop weighs more than a smaller loop causing fatigue to occur more rapidly.

The local Utilities Department contacted us for a metal detector that would find brass "head" valves buried 18" to 36" deep. Here in our community, between every other property line, there are three head valves. Two for potable (drinking) water and one for irrigation (retreated run off / waste) water.

The valves are all attached to PVC pipe so we could expect no help from signals following a galvanized steel or rigid copper pipe to the valve. The valves are supposed to be buried at a specific depth a specified number of feet from the road edge on the property line. The ***actual*** location of the valves are in the general area they were supposed to be ... but not exactly, thus the request for our assistance.

We recommended the Utility Department "invest" in an XLT Spectrum with a 15" loop. We would provide and install a special "notch edited" custom program for their XLT that would discriminate out all metals below and above the VDI (Visual Display Indicator) number range of the valves we needed to find. Our custom program also boosted signal strength to the point where we were getting responses on valves at 32" with no chattering or false signals. Our field test found seven out of seven valves, each in a matter of just a few minutes.

Without the notch discrimination ability that the XLT and DFX offers, the utility crews would have had to dig on targets found outside of the VDI range of the valves. With the larger 15" loop, they are not distracted as much by smaller, shallower targets within the same VDI range of the valves.

The point we make here for beginners, is that if you want to find ***large deep*** targets, go for a bigger loop. If you want to find smaller items extra deep, go for a back-hoe and spread out the soil and use a factory standard loop.

A smaller loop of 3" to 5" inches is quite useful in detecting high trash areas to differentiate between trash and treasure targets that are close together. Smaller loops are as useful to detectorists using mid range priced equipment as they are to the top of the line equipment users.

Mid range visual target identification detectors often assign a target to one of up to about eight target ID zones. If more than one target is under the loop, usually the signal shown on the target ID is from the largest, most shallow target. With a smaller loop you can often read the visual ID differences between close together targets more accurately than you can with a larger loop. It is beneficial to be able to tell the difference between what may be trash versus treasure metals.

If cash flow is available and you think you will need to add a smaller or larger loop to your equipment package, they are useful to have when needed.

We are often asked why most larger loops have a hole in the middle and smaller loops do not ... why some brands feature a "peace symbol" shaped open loop, or a solid elliptical loop, and why some loops are rectangular in shape.

The configuration of a loop being open or closed is the result of the manufacturer designing a long lasting, light weight secure attachment platform of the loop to the lower clevis rod.

The difference between being round or elliptical is for the functionality of the user. Elliptical loops were typically only used in high trash or gold prospecting situations. They generate a longer narrower signal.

Today, many dual and multiple frequency metal detectors use elliptical loops allowing better frequency interactions and supposedly to gain greater depth. Round loops are more commonly used in the majority of detecting circumstances. Rectangular loops can offer greater depth and a wider search pattern.

What is the best type of metal detector for a beginner?
Every type of equipment has a learning curve that takes time and practice to fully understand and use at its optimum level.

Think about how long it took you to learn how to program your VCR or to become a proficient and safe vehicle driver. Don't despair over the VCR example. To me, using a metal detector is much easier than programming a VCR. Perhaps this is because I do more metal detecting than recording of television shows.

Obviously, the higher the level of sophistication of the device, the longer it takes to fully learn how to achieve maximum performance from all the features / user benefits a sophisticated metal detector can offer. Practice makes perfect.

Just because a detector is the top of a manufacturer's line, the most expensive one they offer, and offers the most sophistication, does not necessarily mean it is the most difficult for a beginner to operate. In some instances this is true, in others it is not. Conversely, a manufacturer's least expensive "low end" unit may not be the easiest to learn how to use either. Lower priced units may not provide all the features an adult user will find informative, useful, and enjoyable.

Stay away from the real cheap "no name" metal detectors generally sold by electronics stores known for selling everything to do with anything electrical and the "big box" discount stores that don't offer a full range metal detector department. If the store has no one on staff that practices the hobby, it is not likely you will receive adequate information to make an intelligent best choice for your first machine?

Cheap "no name" units certainly supply a low end beginning price point that is easier on the wallet, but are usually not of the quality to find much more than the most shallow higher phase targets. Cheap "no name" units also tend to be heavier and sometimes more cumbersome to use.

The greatest mistake you can make when buying a metal detector as a gift to someone else, is to purchase the lowest priced almost featureless machine. It is likely such a machine will not be entertaining enough for a child as he or she matures, much less for an adult.

If the least expensive machine is cumbersome and all it is able to find is shallow, larger pieces of trash, the gift will end up not being used more than a few times. Your gift money would be better spent on something else.

Determining value and performance differences between a cheap "no name" detector versus the beginning price point of well respected name brand metal detector can be confusing. At the beginning price point, from name brands, you should expect greater depth, full range discrimination, and a comfortable light weight design. Ask about expected operating time before batteries need replacing based on the type and number of batteries being used.

Battery power correlates to a direct relationship in weight, operating time ***and performance***. A machine that uses a single 9 volt battery is not as likely to perform as well as a machine that requires a 12 volt configuration.

Compare this line of thought to cordless power tools for the home owner providing 6 to 16 volts of power. A professional grade cordless tool requires 24 to 36 volts for heavy duty work over longer periods of use. Now, ask yourself if you'd rather swing a machine for hours on end using 8 AA cells or 8 C cells?

Both provide 12 volts of power. 8 C cells weigh over one pound. 8 AA cells weigh less than 1/2 pound. Name brand machines use and "pull" the required operating power at a constant rate. This means when the batteries need replacing, the power goes down quickly and noticeably. A "no name" machine may operate like a flashlight and slowly, unnoticeably, not perform as it did with fresh batteries. This compares to a bright light beam versus an almost useless dim light beam.

If you select a low end unit that only offers a discrimination setting knob and no pinpointing ability, you will find it very easy to use but will quickly outgrow its sophistication as you gain more experience with the hobby. That means you will be wanting to trade up to a better machine more quickly than anticipated.

An inexpensive metal detector is also more difficult to resell even remotely near its original cost to someone else who has any knowledge of the hobby at all.

If you've ever been a boater you can relate to trading up as you remember going from the used 12 foot "no frills" flat bottom Jon boat and motor to that sleek and super fast "cuddy cabin" cruiser with GPS, Auto Pilot, RADAR, and automatic electric anchor winch. Both will get you to the fishing hole. One will be more enjoyable, faster, safer, more comfortable to use, and perhaps more reliable.

The correct purchase (***investment***) decision is to select the best machine that current cash flow allows and meets the needs of where you plan to do the most detecting. If that means the lowest priced ***name brand unit***, great, go for it! With proper initial instruction and bit of time and practice, you'll soon begin finding enough modern change and jewelry to save until a more sophisticated machine is affordable. If a mid range to top of the line detector is affordable, you may never need to consider upgrading again.

You will need to decide if your primary treasure hunting will be on dry land for coins, jewelry, relics and artifacts, or perhaps for gold nugget prospecting, or for underwater detecting where you and your detector will both get wet from being totally submersed.

The regular VLF detectors are the best all around units for beaches and inland though they may not locate tiny gold flakes (in gold mining areas) so small they would have to be picked up with an eye dropper full of water.

If you plan to do a lot of detecting while SCUBA diving or working the surf area of beaches, you will want to buy a unit made for that purpose. Both VLF single and multiple frequency and pulse induction machines are available for water and underwater treasure hunting.

If you plan to do mostly gold prospecting in fairly clean non trashy remote areas, a detector made for that purpose is what you would find most successful. Gold prospecting units generally do not provide any or very little discrimination of ferrous metal targets and may provide the additional information as to the level of ground mineralization. This can be important to know when you are in a black sand, gold bearing area.

I would recommend a beginner buy the best "general use" VLF detector (that cash flow permits) to enter the hobby. Later, you can always buy the special purpose detector you need based on your interests, experience, and needs.

When selecting the first detector, you can't count on your own experience to guide you as to what machine may be best for you. Obviously cost is an important consideration for most of us but should not be the sole consideration.

From an experience based point of view, the following are the user benefits that contribute to the perfect machine. To me, they are pretty much in order of importance. Full range discrimination is not mentioned as all decent machines offer some form of this.

Signal Penetration Depth: A metal detector should be able to signal the presence of tiny deep "low phase" metal. Small gold / white gold, tiny bits of aluminum foil, etc. Tiny gold / tiny white gold is the most difficult target to detect at depth. Lots of machines can signal on these targets when they are relatively shallow, three inches or less. A deep seeking machine may signal these same targets at depths of eight to ten inches or more with a standard nine to ten inch coil in excellent ground conditions.

Some of this depth will come from transmitting at higher frequencies, some from dual or multiple frequencies with the machines "gain" setting turned up as high as conditions allow with AC and DC sensitivity settings correlated to the gain. Higher frequencies tend to respond better to small gold / small aluminum. Low frequencies seem to respond to higher phase targets like silver deeper and better. If I had to choose a ***single*** frequency detector, I'd opt for one with a higher frequency (approx. 15khz) for the most depth.

Target ID: Very important when deciding when and what to dig. With target ID you can turn discrimination down very low or to almost no discrimination to help achieve maximum depth. Without target ID you have no idea what's in the ground other than it is an item within the phase range above your discrimination setting.

With target ID, you can choose to ignore or dig any signal based on the ID number or symbol. Silver coin shooters can choose to dig only those signals that ID in the range silver coins commonly appear.

Depth Indication: To me, as important as target ID. When hunting for newly lost items like a ring lost in a park after a picnic, festival, or neighbors yard, target depth indication allows you to ignore deep targets.

When hunting for older coins, target depth indication allows you to ignore shallow more modern coins and trash in favor of spending valuable time and energy recovering deeper older coins. Combined with target identification information, you can become hugely more productive with your detecting time.

Weight / Ergonomics: Nothing tires you more rapidly than a poorly designed heavy metal detector. Choose one that is as light in weight as possible and designed to be comfortable using all day long. Machines that are designed to be used underwater as well as on dry land tend to be heavier in weight. Dry land use machines are usually lighter in weight.

When holding the machine ready for action, does it feel like an extension of your arm or does it feel like a heavy tool?

The type and number of batteries used in the metal detector have a direct bearing on the weight. Dependent upon the manufacturer, the type and number of batteries also determines how long the machine can operate at optimum levels before replacements are needed. Another contributing factor to weight is the material used in construction. Aluminum, resin, or plastic each contribute to the machines weight, visual appeal, and durability.

Simplicity Of Operation: A new treasure hunter needs a machine that can be as simple to operate as "turn on and go". A machine with too many analog (switches / knobs) control settings or too many digital control (LCD screen menu) choices to make in preparation for beginning a hunt is not comforting to a new user.

Later on, when you have gained knowledge and experience, the more control settings you have available, the better for you if you want to make a more full range of adjustments for your style and type of metal detecting and ground conditions.

Fortunately, most of the sophisticated detectors made today can be as simple or as complex to use as you make it to be. If the machine you admire offers both "turn on and go" operation as well as all the optional adjustments to transmit / receiving signals and audio preferences, you'll have the best of both worlds for the beginner as well as for the experienced detectorist.

Before buying, ask for a demonstration and review the machines operating manual (s) to judge for yourself how easy the machine will be for you to use.

Tone ID: Not nearly as important as the above features but nice to have. Single tone machines tell you the target is above your discrimination setting. Two tone machines generally tell you if the target is above or below your discrimination setting. Full range tone ID provide tones from the lowest to the highest phase targets above your discrimination setting. If you plan to recover every target above the discrimination setting, tone ID is less in importance.

Top of the line machines offer far more settings / features / user benefits far more complex than what has been discussed above. To need or understand the use of most of these features at this time places you outside the category of a beginner so they are not discussed in this section.

Purchasing a machine that will do everything you may ever need it to do makes sense if cash flow allows.

Ground Mineralization / Ground Tracking

Directly related to the topic "What is the best type of metal detector for a beginner?" is this section on ground mineralization / ground tracking. We don't have a mineralization problem here in SW Florida, so I asked my friend Bob Davis from Republic, Missouri to help explain the topic more intelligently to you. Hopefully, I listened well enough to correctly paraphrase his explanation for you.

Bob has practiced the hobby for over 30 years. His advice and comments on various metal detecting forums is widely respected and valued.

"Actually, metal detector is a misnomer. Your machine is truly a metal ***and*** mineral detector. Minerals are within the detection range of our machines. To prove this, switch your machine to its all metal or pin point mode and sweep it over a sack or block of common salt. This is why the term "hot rock" has become part of our detecting vocabulary. A hot rock is a highly mineralized stone or rock that has no value and causes your detector to respond.

An improperly tuned (adjusted) metal detector may not signal on a coin over three inches deep in mineralized ground while the same coin may be detected over nine inches in ground with low mineralization. When properly adjusted to the ground's mineralization, a metal detector will offer it's peak performance under the conditions present.

Since relatively worthless minerals are of no interest to people looking for metal, our machine must somehow ignore the constant signals received from minerals and respond only to metal. Some machines have a manual ground balancing procedure while others are designed to read and adjust out mineralization automatically many times a minute.

Some parts of the country are relatively mineral free while others are highly mineralized. Ground mineralization can change within just a few feet. The higher the ground mineralization, the more difficult it becomes to achieve optimum depth seeking performance.

Once ground balancing has been achieved, the machine needs to be set or told what sensitivity level (strength of signal) to use before it responds over a target with an audio tone or beep. Set too low, the machine may not respond to good targets. Set too high, the machine may respond with a false signal on targets that aren't actually there. To be successful, a detectorist must insure their machine is stable and operate predictably. This requires the machine to be adjusted properly for the hunting conditions."

Being new to the hobby, and if you live in an area that may have varying ground conditions, you may want to purchase a machine that offers automatic ground balancing so you are sure it is adjusted correctly. Ask your local dealer about how area ground mineralization may affect your success with the machine you are wanting to purchase. If he or she practices the hobby, you will get good advice.

Buying Versus Renting

We are often asked, "Can I rent a detector for the day to see if I'm going to enjoy the hobby?"

Most detector stores rent one or more types of detectors. The down side of renting is that it is expensive at $20 to $40 per day on average. It will almost always require several trips to different types of treasure hunting sites to begin learning about and truly enjoying the hobby. People aren't likely to rent a metal detector more than once to find out if they will enjoy the hobby.

Most detector stores discount their machines for sale at prices where it makes more economic sense to buy a unit and use it for about a year. If you find the hobby to be less enjoyable than you thought, discount it for the amount of renting it for two or three days and sell it to someone else. Buying a metal detector can be less expensive for you than renting a metal detector for two or three days … and you will have use of the detector for a year or more!

When buying a metal detector, find out if the unused portion of the warranty would be transferable to the new owner should it be resold.

How can I find out if I'll enjoy the hobby?

We think the best way is to purchase a metal detector and start detecting on beaches, parks, athletic fields, and old home sites.

The second best way is to locate the nearest metal detecting club and go to a couple meetings to observe, learn, and meet some experienced metal detectorists. You'll almost certainly be able to meet someone who may have an extra machine and invite you to tag along on a trip or two to see if your interest level continues.

Another option is to visit your local metal detector dealer and express your interest in getting into the hobby. Ask to be introduced to someone who may take you under their wing for a trip or two and teach you the basics of the hobby and perhaps give you some direction as to what machine may be right for you to purchase.

Where should I buy my first metal detector?
Your first metal detector should be purchased from a local retail store where at least one complete line of detecting equipment is offered … ***if*** someone on staff actually practices the hobby themselves. That way, you'll get experience based answers to your questions *before* you buy and more importantly, initial training. You'll receive training not only in your machine's operations, but in detecting and recovery techniques as well.

There are many retail and Internet sources to buy a detector. Some of the retail outlets will only have a few detectors from which to choose and may not have anyone on staff that can give you correct information as to which unit is best for you or how to operate it effectively.

Some Internet sellers are better than others. You will almost always be told their prices can not be beat. You will be told their selection is the best quality money can buy and they will quickly offer to take your credit card number over the phone.

A savvy first time metal detector buyer should make a point to capitalize on the experience and knowledge available from a dealer who participates regularly in the hobby. It is important that you can test the detector you think you want right at the store … before you buy it!

A few times we have sold metal detectors to people who had bought their first metal detector from a "big box" department store or mail order house. They got frustrated at problems they were having and had no one to quickly go to for "hands on" demos and answers. In a couple instances we were able to determine that a detector bought somewhere else had been defective right out of the box. The user, as well as the seller, wasn't familiar enough with the machines operating capability to realize it was not operating properly when it was sold!

Detector dealers are smart enough to quickly determine if you bought your machine from them or not should you bring one in to have a whole host of questions answered or potential problems with your detector resolved.

All successful metal detector dealers will try to help you to the best of their ability. If they are busy when you come in, who do you think will get the most time, thought, and attention to your needs, the dealers customer, or a competitors customer?

One of my detecting friends sent a friend of his over to see me because he wanted to get into the hobby and didn't know what machine to buy. We spent about 1 1/2 hours together over two different visits. I really thought he would come back and buy an XLT.

I later heard from my detecting buddy that his friend had bought a new XLT from a distant catalog shop where he saved almost $10 after freight was considered ... but was frustrated in that he couldn't get his technical questions answered.

A few days later, I received a call from our White's distributor saying a guy called with a lot of questions on a brand new XLT he had recently purchased. The distributor advised the gentleman to call me to get all his questions answered. Our distributor told him there were two places in the general area that sold Whites and my store was the only one where he could get all the experienced help he needed with his new XLT.

The man said he didn't feel right about calling us "because they spent so much time with me and I ended up buying somewhere else to save a little money. I'd feel bad about calling."

I laugh because we have caller ID on our store phone and can tell when the gentleman does in fact call to ask his questions on the XLT and about detecting places in our area. I don't want to embarrass him. As long as he keeps trying to act sneaky, I'll keep answering his questions ...when I'm not taking time away from one of ***our*** customers.

Later, when you become more experienced and are wanting to upgrade or perhaps buy an additional special purpose detector, check out the price, service, and reputation of all those who sell what you want.

Don't assume the local dealer where you bought your first detector won't be able to competitively serve your needs again. When the local dealer is competitive in price (and most of us are very competitive), buy it locally. If the dealer is way out of line in price for the same "apples to apples" brand and model number machine, you should be encouraged to go elsewhere. Remember, from legitimate service oriented sellers, you always get what you pay for.

Successful metal detector dealers know their product, can answer all your questions honestly, will have competitive prices, can give you good detecting leads for local hunting and will be around for you should you need their help. They will not survive long when their prices are out of line (too high or low)!

All the Internet metal detector user forums experience way too many postings of very basic questions that should have been quickly and easily answered from where the detector was first purchased, regardless of the machine brand in question. The same questions are asked and answered time and time again to the frustration of those who are seeking more advanced help on the forums.

Another important reason to buy locally (for everything you purchase), is to help keep your business community vibrant and prosperous. Locally owned businesses contribute to the local tax base, support local charities, and spend their profits employing your friends / neighbors.

To help insure you can continue having a choice of where to shop, ... buy locally! "Big box" retailers and Internet sellers spend their profits outside of your area contributing little or nothing to your community.

What You Can Expect To Find

Trash, ... lots and lots of trash with occasional coins, artifacts, and jewelry mixed in among the trash. Remember, the treasure hunter that digs the most trash will find the most treasure!

If it is made from metal, a decent detector will find it. Since the bronze age, people have lost or discarded metal items you can find. This is a good news / bad news situation.

The good news is some old trash may be considered treasure today. A *broken* ancient Greek or Roman broach was discarded ... thrown away by its owner. When found today, it could be valued up to several thousand dollars depending on its rarity and condition. A *lost, unbroken* broach from the same era could be considered priceless and perhaps a national treasure.

A 1800's dented button, broken military insignia, fired or dropped military round, old watch, or watch fob, could have been yesterdays trash and today's treasure.

The bad news is that even some ancient trash will remain trash if it has decomposed to the point where it is unrecognizable. The bad news gets worse when you realize that people have been littering the world since man first began making things that break or wear out. Unfortunately, this has continued right up through to modern times. Trash far outweighs the good stuff lost and recovered.

The really good news is for coin hunters all over the world. Coins became the first form of payment outside of trade goods beginning in the pre Christ Celtic period somewhere about 150 BC.

In 1997, the U.S. Mint estimated there were about 16 billion U.S. coins actually in circulation. That number would be somewhat higher today to supply a growing population, say, perhaps 25 billion. Every year huge numbers of additional coins are minted. In year 2000, 28 billion were minted. The most ever made in the U.S.A. in one year.

The mint estimates it takes about 25 years for a coin to wear out. If they last that long, why are so many more made every year? Two factors take coins out of circulation. Everyone has coins stored away somewhere outside of their pockets. Children fill piggy banks with them. Adults fill jars or coffee cans with accumulated unused change at the end of the day to someday be "cashed in" for paper currency.

Joining these coin savers are coin collectors who remove coins from circulation to complete their collections. Some day these new coins, worth face value, become old coins worth much more. Ok, this accounts for some of the coins removed from circulation, what about the rest?

Most are just lost. They drop from hands when making change. They may come out of a pocket entangled in with a set of keys. They fall from loose pockets or pockets with holes.

In the days of less law, no banks, and no secure places to sleep at night, gold and silver coins were hidden anywhere a person found handy near their campsite or home.

Here in the U.S.A., "coin shooters" concentrate on finding pre 1965 coins because then, the silver coins were actual silver, not silver plated clad base metal. Gold coins were once commonly circulated and on rare occasion, will be found. Europe and Asia are a coin shooter's dream location for finding really old and valuable coins.

My detecting friend Norm, always begins every search by saying, "It's out there, all we have to do is find it." I don't know if he's that good ...or just that lucky, but he does find a lot of interesting items.

By working parks, athletic fields, schoolyards, and beaches on lakes and the seashore, I have found as little as $250 to almost $1,000 a year in modern US coins. I try to go out once a week for three or four hours of detecting either for myself or in helping others to recover their lost items.

Once I picked up a crumpled "empty" cigarette pack from the beach to find a $100 bill tucked inside it that was probably thrown away by a beach visitor that should have had one less adult beverage. That certainly was a piece of trash that turned out as treasure.

In our area of SW Florida we head to the beaches more often than anywhere else due to its pleasant environment and our chances of finding lots of coins and jewelry is better there.

We don't have the history of being populated with a lot of large and small towns since the beginning of our nations history like they do in the Northeast and Midwest part of the country where older coins and relics are found in yards, parks, and any of the older home sites … in between the pieces of modern trash.

My friends and I average finding a ring or pendent that we can keep about every seven trips to the dry sandy beach. When detecting in the water, one of us almost always find a piece of jewelry. The colder water, fast currents, and higher tides found on the Atlantic and Pacific Coasts contribute to more people losing more jewelry there than here in SW Florida.

If you like to travel, make sure to take along the detector even if you don't think you'll have the time or know where to go. You'll be surprised at the opportunities you'll find. If you like going on cruises that stop at exotic beach islands or travel to other cities on a visit, take along the detector.

If you stay at Bed and Breakfast facilities, sometimes you can receive permission to search their property. If people have been there before you, they have lost something you can find.

Initial Tips For New Treasure Hunters

OK, you have your new machine and now you want to use it effectively. Here are a few ideas to consider to make your learning go easier.

Adjust the length of the down rod (s) of your metal detector and the angle of the loop to the down rod so the loop is parallel to the ground and your upper arm is vertical to the ground when in a resting stance. See the photo to the right.

You want the loop to extend out far enough so that it will not be in the way of your feet when slowly walking. You will tend to walk faster than you should if the loop is extended further out on a longer down rod.

Sweep the loop from side to side stopping the swing slightly past where your left foot would be if you are swinging the detector in your right hand. This would be opposite for left handed people. If you sweep all the way across your body you will become fatigued too quickly.

Try to keep the loop parallel to the ground surface, as close to the ground surface as possible, without actually "kissing" or scuffing the surface. Learn to slightly overlap the loops path as you swing it from side to side to insure the area to be detected is searched as near 100% as possible.

New treasure hunters have a tendency to not keep the loop parallel to the ground and to raise the loop too high off the ground. They also tend to sweep the loop up into the air at each end of the swing. New treasure hunters need to practice keeping the loop parallel to the ground throughout the entire swing, start to end.

Remember, generally speaking, the loops signal will penetrate the ground a little deeper than the loop is big. If the loop is held one inch off the ground's surface, you are only giving up one inch of treasure seeking depth. A loop held six inches off the ground will mean you are losing six inches of your machine's treasure seeking ability and possibly missing many great finds!

When you go too fast you'll miss searching the area between each loops swing if you are not overlapping on each swing. The frustrating part of the hobby for all treasure hunters is too much land to cover with a limited amount of time with an average loop size of 8" to 10". All of us have a tendency to go too fast. It seems the slower and more carefully you search an area, the higher the quantity and quality your finds will become.

The speed at which you swing your loop varies with the type of detector you've purchased. Practice swinging the loop at the speed recommended by the manufacturer. Lower priced units may require a slower sweep speed than higher priced units that operate effectively at faster loop speeds. Even top of the line machines will provide better depth and target ID when you slow down over an "iffy" signal received while going faster.

Discrimination Setting Tips

The discrimination setting on your metal detector will tell it what type of metal to ignore. Discrimination can be the most helpful or hurtful adjustment you can make on most metal detectors.

In a high trash area, too low of a discrimination setting will drive you crazy in responding to and recovering signals. Too high of a discrimination setting will keep you from finding smaller gold coins or jewelry. The more discrimination you use, the more depth you give up. Even though the depth given up to higher discrimination isn't major, you never know if that old coin or relic was not able to be found due to a minor depth disadvantage.

When detecting a high trash area like an abandoned old home / plantation site or ghost town, that you know has potential for recovering worthwhile items, you need to plan accordingly. The plan might include more than one trip with different discrimination objectives on each trip depending on the size of the site.

The first objective would be to locate initial hot spots, or areas within the site where desired items may seem to be concentrated. Areas that might be hot spots would be around larger older trees, where the hitching post may have been located, where the porch may have been located, where the outhouse was in relation to the main house, slave quarters, evidence of the cellar location, a well, etc.

To locate these hot spots, set the discrimination on your detector where you can pick up copper pennies and silver dimes or better and ignore small aluminum and ferrous targets for the first search. At this setting you should not receive signals on small shards of cut up aluminum cans, pull tabs, nickels, and small gold items.

When you start finding copper and silver coins or mid to large gold items in a hot spot area, turn discrimination down to where you aren't getting ferrous target responses and small pieces of aluminum foil but will pick up nickels, pull tabs, small gold and higher.

Work the hot spot area thoroughly removing all trash from the area. Repeat this process until all the hot spots on the entire site have been searched to your satisfaction.

Relying too much on discrimination or target ID will keep you from finding many valuable pieces of jewelry and old coins / relics. Discrimination allows you to more quickly "survey" a site for evaluation of potential finds and more quickly locate hot spots.

Recovering a target based solely on target ID is also a form of discrimination. Relying too heavily on target ID will may keep you from recovering valuable targets. Be aware that a targets size, position in the ground, nearness to another metal object, or depth may change its expected target ID reference number / indicator.

Here's an example. In parks, yards, athletic fields, and beaches, lots of gold and silver jewelry and coins are lost. One day when working a soccer field, I decided to recover all targets in the normal "ring range" or higher regardless of depth, and all "foil range" targets deeper than two inches. I was surprised to learn that small gold items (chains / tiny rings) would sometimes ID in the foil range at depth, but ID in their normal ID range when placed on the surface.

The reason I ignore shallow (less than 2 inches) foil is that foil is light in weight and typically remains shallow unless buried by some other means such as from fill dirt being brought in or from tilling.

One of the Internet forums had a posting from someone who finds gold and silver jewelry on the same football field almost every time he visits. He attributes his success to using no discrimination and setting his goal of slowly searching "every square inch" of the limited amount of ground he has time to search that trip. Some days, his goal may be an area of just 10 feet by 10 feet square. He recovers every target. The next trip he resumes his search where he quit the last time.

It took him over a year to cover the entire field the first time. Now that he's removed all metal targets from prior trips, he can cover more ground digging less trash targets and finding more coins and jewelry on each future visit. This approach makes a great deal of sense if you have unlimited permission and time to work a site with good potential.

Where Can Beginners Treasure Hunt

Start in your own yard. Learn your machine's functions by trying to determine trash (construction debris) from treasure. If you live in an old home you are far more fortunate than if you have a newly built home on land that was not inhabited before. Older homes have had more people over the years losing things than new homes unless your new home was built on land that was in public or military use over the past hundred years or so.

Start along your driveway and look for coins dropped while loading and unloading from moving or from getting in and out of vehicles. If there are old big trees in your yard this is an excellent place to look for items dropped while someone was sitting or laying down enjoying the shade on a hot summer day.

If you home is so old that it had an outhouse at one time, determine where the outhouse was and detect a path from the back door to the outhouse site. Where were clothes hung out to dry over the years? The area under clotheslines almost certainly have a coin or two waiting to be found. Does your home have a porch or crawl space with dirt under it that you can get to? If so, check it out. You never know what someone dropped there or some child took under the porch to play with and lose years ago.

Are there ponds or lakes nearby that have a swimming / sunning area or beach? Watch the water level for it to drop enough to reveal a bank that months before was under water. This is a great time to check out the boat launches too.

Are you a member of a community swimming pool association? In early fall and late spring before or after the pool is open for business and seasonal maintenance is being performed, with permission, you are very likely to find coins and jewelry in the grass areas where people relax when not swimming.

I once detected around an old public pool in north Georgia, no longer in use, that had been filled up with soil to the edge of the walkway around the pool. Within the first ten minutes I recovered two rings and over a dollar in coins.

Head for the city park very early in the morning or very late in the afternoon / early evening when it's not being used by a bunch of children. Check out the sandy areas under the gym equipment and under the swinging tire swings. Also the picnic areas. Please fill in all holes you make recovering targets!

When the carnival comes to town be there the morning after tear down to recover the coins and jewelry dropped into the grass from the vendor booths and rides.

Does your church have or has it had outdoor grassy area activity from pot luck dinners, revivals, festivals, or child care / education? Get with the Minister for permission to check it out.

Does your town have outdoor festivals celebrating an historical event or famous person sometime each year? These areas are loaded with dropped items and trash you can find with your metal detector.

How about the local "lover's lane" areas in your community? You never know what gets dropped and lost while watching the nightly "submarine races". These sites do tend to have a lot of trash but also a lot of treasure if you find the right place at the right time.

The high school or community football, baseball, and soccer fields (sports complexes) have provided me with hours of detecting entertainment.

Do you live in a part of the country where gold mining was done in the past? There are still nuggets to be found for those brave enough to face the trash that can accumulate at some of the less remote "club hunting" sites. Look for the most remote gold mining areas as they have less trash. Hunt the stream banks where soil is exposed and the gullies where rain water has eroded hill and mountain sides.

How about old logging or railroad camps, church camps, Boy / Girl Scout camps? How about land where an old school was once located?

Ghost towns are not all out west! Check out the old maps at the City Hall or County Seat. Compare them to today's maps looking for roads that used to be there but no longer are. Look for small towns on the old maps that today's maps may show farm, field, or pasture. Look on the old maps for where military installations were once located. Make same size transparent overlays to put over the most current maps.

Read the history of the Civil War in your area. Some "marches" were actually running battles fought within a few miles of both sides of the road or railroads between towns.

How about "the old swimming holes" that were commonly used before swimming pools became popular. These once popular areas can usually be found near roads and bridges. Ask the local old timers where they used to swim as teens or young adults.

Why And Where People Hide Treasure

I can remember as a young boy of less than 10 years old, the lumps under the carpet in my grandparents' bedroom. Once after tripping over one of the lumps, I decided to discover what was causing it. I flipped back the carpet edge to find layers of paper money going back under the carpet as far as I could lift it up. In retrospect, that currency must have been worth many thousands of dollars. It was "hidden" there by my grandfather who never trusted banks since losing all that he owned during the stock market crash and the great depression.

I knew the money under the carpet was something I should not be playing with so I put the carpet back down and never looked again.

"Big Ed" died in the early 1970's and his edge of downtown home property was eventually sold to a church. As I think back, I have to wonder what must also have been buried in his back yard or under the house in the dirt crawlspace! Were more bundles of cash buried in the yard or under the house in a metal container? Were there gold coins secreted under the soil of the relatively small back yard? Perhaps someday I'll get back to that state and town and be able to receive permission to detect this old family property.

Years ago, Australia passed legislation banning private gun ownership. I talked with a tourist to our area from Australia who told me right after the law went into effect, you could not find any 4" PVC pipe, threaded caps, and female thread adaptors in stock at hardware stores in the country. When I asked why, he told me it was all bought up by people burying their firearms to keep them from being found and confiscated by the Australian authorities.

My guess is that there are many hundreds of thousands of sites in our country and around the world where people have buried money or treasure to be recovered later as it was needed.

If you live in a part of the country where the Civil War was fought, you might have sites where household gold, silver, and personal jewelry was hidden away from advancing Union or Confederate units as they fought their way to their objectives.

During prohibition many millions of dollars were made by those smuggling or making their own beer and liquor and selling it illegally. In order to hide those ill gotten gains from the tax man, much of the money was buried. Find newspaper archive records of these people for clues as to where they worked or lived. These may be good sites to search with permission of the property owner.

Almost every part of the country has stories of highwaymen or other criminals who have robbed stagecoaches, trains, stores, and banks and secreted the loot before they died in a hail of gunfire or were caught. Are you willing to spend the time and energy to research and find their hidden treasures?

Like my grandfather, there are many people who did not trust banks or government with the wealth they accumulated nor did they think it was the government's business as to how much they had accumulated. Every old home site might be a source for hidden treasure or worthwhile relics and coins.

Don't think that all that is worthy of being found is buried. I've read of rare and valuable coins being hidden well above the ground behind rocks used in construction, in mortar joints, behind removable chimney stones or mantels of old home sites, stagecoach stations, taverns, and other public and private buildings. It's always worth a minute or two to sweep the loop over these areas.

You sometimes hear of rare coins or old weapons being found when old houses or public buildings are demolished. A thin coat of plaster has hidden treasures for years. Can your machine discriminate between nails, gold / silver, and antique weapons?

When traveling through your area on trash pick up day, look to condo and hotel / motel trash pick up areas for old vacuum cleaners being thrown away. Remove the old vacuum bag, seal the vacuum bag in a trash bag (in case it has fleas or other nasty stuff). When you get home, sweep the bag with your detector for lost jewelry signals.

You'd be surprised how easily jewelry gets lost in carpet and then picked up by a vacuum cleaner. I know of a man who has a housekeeping supervisor from a large vacation rental condo complex save all full vacuum cleaner bags for him rather than just throwing them away. The supervisor must think he's nuts to want old nasty vacuum bags.

How To Find Worthwhile Detecting Sites

Ask other treasure hunters. Sometimes they may be helpful, other times they will not because they want to keep their better sites for their own exclusive use.

Detecting Clubs can be helpful when all members are required to make a at least one new site available to all other club members each year. Clubs that don't have this requirement tend to have only a couple members that do all the lengthy research to find good sites. All the other members want to detect upon it without making their own site contribution.

Do not believe others when they tell you there is nothing left to find on a site that has been heavily worked in the past. Certainly there will not be as many items to find, but there will almost always be a few!

I was once invited to detect a Federal Seminole Indian War fort site mid-way between Cape Coral and Orlando that had been heavily worked for several years by four or five experienced detectorists. The site was a pasture across the road from a city park at the edge of the small town. They told me this was probably going to be their last trip as they were not finding much of anything worthwhile any-more.

Two of the guys made a beeline over to an area under a big tree where they had found during their last visit some 1830's coins and military buttons. They swept their loops over and around the filled holes from the week before where one of the buttons was found. They seemed to work that small area quite professionally and completely. They received no signals or at least didn't decide to dig on the one's they were receiving so they moved on with their top of the line metal detectors to other parts of the pasture.

I followed behind them by about 20 yards and went over the area they had just swept. Not five inches from the hole they had filled last week where the button was found, and where they had just left, my XLT responded with a solid signal with VDI numbers indicating small brass and then pinpointed the target to be six inches deep.

I carefully recovered a Georgia State Militia uniform button. Being relatively new to relic hunting at the time, I did not realize what I had found other than a brass button with its loop still intact on the back and the State of Georgia Seal clearly evident on the front.

My buddy Jerry saw it and almost went ballistic with excitement. He said very loudly to me, "Do you know what you have here? Do you have any idea what it's worth?"

I said "No, but make me an offer!" He grinned from ear to ear and said, "I'll give you $250 right here on the spot if you'll take it now."

Knowing my friend, I told him I'd pass on the offer. Before we got home that evening, the offer was up to $500. I still have the button and two standing offers for it.

We can't explain why a Georgia Militia button would be there unless the Federal troops had requested help from the Georgia Militia to fight the Seminoles. It is the same button worn by the Georgia Militia during the Civil War. We have not found record of any Civil War activity involving the Georgia State Militia for that small town. Go figure?

Later trips to that "worked out site" by those same guys revealed a brass military spur, more buttons, and more old coins including a $20 gold piece. If the site is ever plowed, I'm planning to spend as much time there as possible.

Deep plowed farm fields tend to become more productive about every third year. The plows turn the soil to bring "new" old targets up to detectable range. Don't give up on fields that you found stuff on a few years ago.

It is almost impossible to assure that every square inch of ground in an area to be searched is adequately swept by a detectors loop. Here's why. Near the surface, a cone shaped signal is at its widest point. The deepest portion of the signal will be its narrowest point. Sometimes the deepest point of the signal may be little more than an inch or so wide. This is why older sites may never be fully worked out.

To fully search an area at a round loop's maximum target seeking depth, would take 8 to 10 sweeps per foot of forward distance to be completely searched.

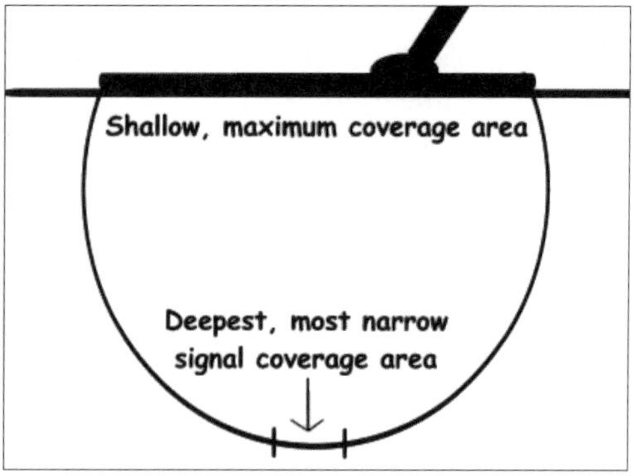

Very few detectorists will be that patient. The really great ones are! Great detectorists are an outstanding example of the old 80/20 principle. 20% of detectorists are patient enough to work an area carefully resulting in them finding 80% of the good targets found.

A slower sweep speed may provide a weak signal at the edge of the machine's deepest signal range. A faster sweep speed may not allow your machine to respond to deeper "iffy" targets at the edge of the signal. Patience is a metal detecting virtue as well.

You and your detecting friends will get distracted from your pattern by recovering targets, talking, showing each other nice items you've found, coming back to detect after breaks, and as fatigue sets in after walking and digging all day. These are reasons why no site is ever fully worked out!

A well worked popular detecting area will also have less distracting trash targets still in the ground as most have been removed by prior treasure hunters. Signals you receive are more likely to be good.

Test Garden

A great way to learn how to get the most out of your new detector is to set up a test garden burying known targets at known depths. Practice finding these targets with your detector adjusted differently as to sensitivity, discrimination, loop speed, and any other signal element adjustments your detector may allow.

First, set your detector to its "all metal" mode (no discrimination) so that all metal in the ground will cause a signal. Find and remove all signal causing metals from the test garden area.

Bury the same test object at 4" and 10". In some areas, ground mineralization will reduce your machines depth seeking potential to 8" or less on a quarter coin, and even less on a smaller coin.

Expect less detector performance from an object buried in the ground versus the same object's signal in an air test. An air test may indicate your machines maximum depth potential. The test garden will allow you to learn how to maximize your machines potential under actual conditions. That's the purpose of a test garden.

Bury your objects in a row / column pattern with each object at least 18" to 24" apart to insure your loop is reading only one target at a time.

Bury items like pennies, nickels, dimes, quarters, brass keys, silver rings, gold rings (precious metal test targets can be purchased inexpensively in a pawn shop), aluminum cans, bottle tops, nails, large and small pull tabs, lead sinkers of different sizes (simulates musket / mini balls), and brass buttons .

At the end of one row or column, bury a trash target and a good target at the same depth but almost touching each other. This will test your ability to pinpoint and determine if your machine can tell you if more than one target is close to another, one being junk the other desirable.

If you live in an area where you run into different ground conditions that may cause your machine to "false" or chatter, you may want to create a "mobile" test device. Use paint paddles can be obtained from your local hardware or paint store.

Glue common targets you want to find on the paint paddles. If you think your machine isn't going as deep as you think it should, or if your machine does not seem as sensitive as it should in difficult ground conditions, cut a slice into the ground at an angle with your digging tool to the depth equaling the common depth of targets found in the area.

Slide the paint paddle into the slice you've made and then test / reset your detector based on the known target at the known depth.

Target Recovery, Beach

Ground conditions vary greatly from soft fine sand to dense, dry, hard as concrete soil. Different techniques and equipment will be necessary to quickly and effectively recover targets without damage under different conditions. Soft sugar sand beaches are perhaps the easiest from which to recover targets.

You can be more productive in recovering targets with a sand scoop. Most scoops have a sturdy handle and digging / scooping

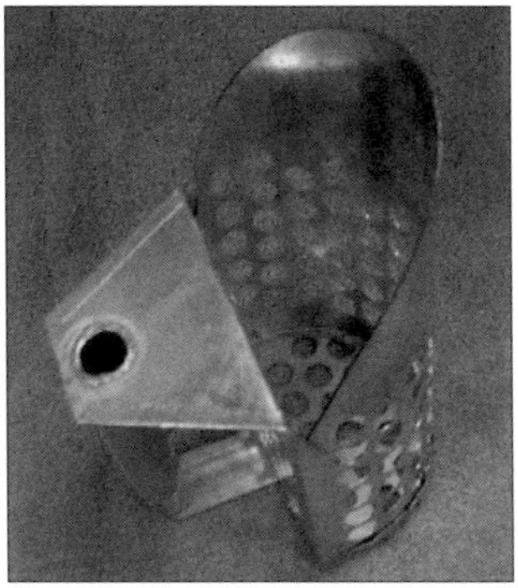

snout with holes in the body of the scoop to allow sand and small shell fragments to fall out as it is shaken.

Some scoops have a hand grip handle, others have a longer shaft type handle that allows for a golf club type swing to scoop out the target with sand from the hole.

Scoop up sand from where your detector has indicated during pinpointing. Sweep the loop over the dig hole to see if the target, or perhaps another target is still in the hole. Don't assume your scoop has no treasure in it if you are still receiving a signal from the hole you've scooped. There may be more than one target!

Empty the scoop on the ground about a foot away from the hole and take a quick sweep over the sand you've just emptied from the scoop. No signal, go back to the hole for another scoop. If there is a signal in the discarded sand, try to find the target visually and / or scoop up some of the discarded sand and shake it out less aggressively to hopefully leave the target in the scoop.

Sometimes the target will be small enough to fall through the holes in the scoop and you'll have to process the discarded sand by hand.

To do this, remove all jewelry from your hand and arm. Pick up some sand from the target signal area with your hand and pass your hand in front of your loop. If you get a signal, the target is still in your hand. Pour off some more sand into your other hand then pass the original hand in front of the loop again. Keep doing this until you have no sand and only the target remaining in your hand. Don't forget to sweep the scoop hole and your discard pile before you fill in the hole. There may be other targets yet to recover.

Here is another recovery technique for finding those tiny targets that fall through the holes in your beach sand scoop. Scoop up sand from the "shake" pile. Check the shake pile with your detector to make sure the target is in the scoop and not in the shake pile.

Place the loop of your detector over the shake pile about six inches to a foot up off the sand. Slowly and gently shake the sand from the scoop so that the sand falls on the loop.

Keep your eyes focused on the loop as you are gently shaking the scoop. As the tiny target hits the loop, or falls past the loop, you will hear a signal and will often see the target on the loop or the dimple in the sand where the target re-entered when passing by or through the loop.

When detecting a beach littered with tiny iron / steel fragments, epoxy a ceramic magnet into the bottom of the sand scoop to attract any tiny pieces of ferrous metals you may recover while treasure hunting. Make sure to epoxy the magnet with the ferrous metal attracting side out.

Some beach treasure hunters will place a screen of 1/4" galvanized "hardware cloth" inside the scoop to keep small targets from falling out of the scoops holes. This is a good idea when water detecting but kind of a pain when dry beach detecting as it takes longer to process a scoop full of sand and small shells.

Wear a two pocket nail apron or some type of soft pouch that has two pockets. Use one pocket for trash, the other for treasure. Always take the trash you find off the beach with you for proper disposal later. It's just the right thing to do and you won't have to spend time finding the same trash targets the next time you detect the area.

Please, fill in all holes you dig. Even though there are children on the beach digging, making sand castles, and leaving holes large enough to bury a sport utility vehicle. You and your fellow treasure hunters will be criticized for leaving the tiniest trace of ever having set foot on the beach!

Wet sand at the edge of water, and sand under water is tougher to work. Wet sand at the edge of water is heavier and more dense and generally requires a heavier duty long handle scoop or short D-handled narrow blade shovel to quickly recover targets.

When working the moist / wet sand at the edge of water, recover the target and empty the scoop or shovel away from the hole you've dug putting the discard sand up hill from the water's edge. Sweep your foot through the discard pile to spread out and thin the sand from the scoop or shovel and visually check for the target. Sweep the loop over the discard pile as well as the target recovery hole until all targets are found and recovered. Attempt to replace the scooped out sand back into the hole to leave it as you found it, less the target.

Recovering targets in the surf is about as tough as it gets! Hopefully you'll not be working this type of target area unless you have a 100% waterproof detector.

A long handle heavy duty scoop designed for recovering targets in water is a must in working surf areas. A shovel will not work. The idea is to find the target and scoop up sand and shake / wash it out until the target is recovered from the scoop. In surf, keep your back to the surf and be careful not to be knocked down or be in the way of swimmers and boaters.

Serious surf treasure hunters will have a 1/8" to 1/4" nylon or galvanized wire mesh lined float in which to quickly dump the sand scoops contents. Natural water / wave action washes the sand from the contents through the mesh, leaving only larger shells, trash, and treasure in the mesh. This saves time and energy of "shake washing" each individual scoop full of sand in the process in recovering the signals detected. This can be important when detecting at night or during competitive water hunts.

When water detecting in rivers, lakes, or ponds, realize the holes you make in scooping up targets may create a danger to non-swimming waders who may fall and drown from stumbling into your recovery hole. ***Please fill in your recovery holes*** by dragging in surrounding bottom material with your feet and / or scoop. When working beach surf areas, this is rarely a problem as tide and wave action fills them in for you in just a few minutes.

For your own peace of mind and for the safety of others, it is best to not water detect when people start playing and swimming in the areas you are detecting. We hear of instances where children get too close and step on the bucket of the long handle water scoop causing the child to receive a cut on their foot.

Try explaining to an angry parent that their dear little (now bleeding and crying) child was being a royal brat and was warned a half dozen times why it is important to stay out of your way. Detect early or late to avoid the busy times.

We are often asked if we recover more targets from the soft dry beach sand or out in the water. The answer to this depends on many variables. Here in southwest Florida our water stays fairly warm, we have very little current and much less severe tide rise and fall with almost no wave action at all. What this means is that there are not as many jewelry items lost out in the surf due to wave action and cold shrunken fingers. It does happen, but not as often as it does on the Atlantic and Pacific coasts where water is colder and wave / tidal action stronger.

If you concentrate your time and efforts in the water, that's where you'll recover the most treasure. The same is true for the soft dry beach sand areas.

Consider this. As a percentage of the population of beach users, treasure hunters are a very small, even tiny percentage. Out of us that are treasure hunters, an even smaller number of us regularly work out in the water. Less detecting competition for available targets!

The old 80 / 20 principle again applies in that 20% of the treasure hunters develop the skills and determination that results in making 80% of the finds. Work slowly and carefully in a pattern.

Target Recovery Inland

It makes a difference in the recovery tools needed depending on if you are working a weed covered or plowed field or a well manicured lawn in a park or yard. Regardless, make sure that when you walk away from a target recovery, you've repaired all signs of having dug there.

In fields where you do not have to worry about turf damage, you can use a short D-handled shovel or folding Army type shovel. Just make sure you re-fill the hole to make it look as near to how you found it as possible.

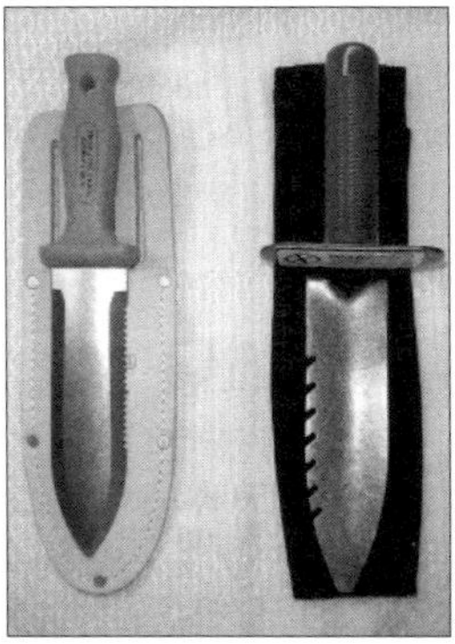

In lawns, you will need to learn how to more accurately pinpoint the target before recovery. In recovering the target, a sharp bladed combination trowel / knife is useful in making a small slice or cutting a small three sided plug in the ground just outside the area you determine the target is hiding.

Some hunters insert a probe such as a dull tip brass rod with a ceramic or solid brass handle to "feel" the target before digging any further. With experience, you can tell the difference in the "tick" received between a metal object and a stone.

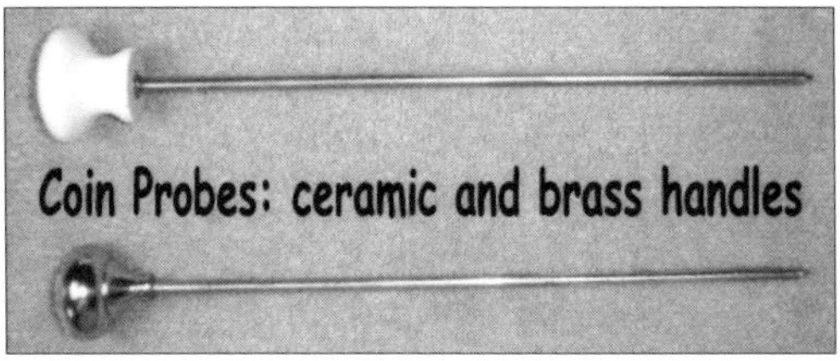

If your ground has a high ratio of stones and gravel to soil, a probe will prove almost useless. Nothing will ruin the value of an old coin or relic much faster than scaring the item with your digging blade.

Beginners will be more successful in using a combination trowel / knife to cut a small three sided plug and fold the plug back out of the way on its fourth uncut side. Spread out a digging towel like a shop towel or thin piece of cloth beside the hole. Process the dug dirt on the digging towel until the target is recovered.

Use the towel to pour the processed dirt back into the hole. Fold the plug back into the hole pressing it back into place with your hand, then your foot. There should be very little to no loose soil left on the ground where you were recovering the target when you use a digging towel.

When recovering targets from well kept lawns, carry a canteen of water with a soluble fertilizer in it to pour on the plug after it has been tamped back into place. This quick treatment can help to keep the cut plug from turning brown and wilting as easily until sprinklers give it further needed care.

Avoid detecting public parks, athletic fields, and well manicured residential lawns during dry hot weather when target recovery will likely leave a dead dry spot in the turf.

On occasion, in the heat of the dry summer, I've had to contend with soil so hard that a trowel or knife would not penetrate the ground more than a quarter of an inch at a time. To dig a target quickly with less effort required a small hand pick or miner's pick. Obviously this technique will cause damage to a valuable target if hit by the pick.

Should you encounter "brick hard" ground conditions regularly during the most dry part of the treasure hunting season, you might want to invest in a heavy duty cordless electric drill and use a 3" or larger hole saw and bit extension to cut plugs deeper and deeper until the hard crust is penetrated and the target is recovered.

Most brands of cordless drills have belt holsters available that make carrying the drill much easier. Since a good hole saw bit will be rated "bi-metal" (to cut through wood and metals) and has sharp teeth, practice your pinpointing skills so you are less likely to damage a target with the hole saw.

Beware Of Detecting Pranks

Sometimes mean spirited kids or adults will try to "prank" you by tossing coins in the water, on the beach, or in a grassy park ahead of you and watch you work for the penny's thrown. The best thing to do here is move off to a lesser used area quickly as possible.

Perhaps you'll be asked by a prankster to find a non-existent jewelry item that was never lost. A good way to combat this is to tell the person you are hunting for another item lost by a friend and that you have limited time to search. Ask the person to write down their name, address, and phone number along with a detailed description of the lost item. Tell them if you should find it you'll give them a call. If the item was in fact lost, they will give you all possible information.

Here's how to defeat the "spotters and claimers" scam. Unfortunately, there are folks who have lots of experience in observing and trying to take advantage of new detectorists. The spotters and claimers will work in pairs. The "spotter", a nice looking man or woman will come up to you and try to start a conversation with the ultimate goal of convincing you to show them any nice jewelry you've found. Don't fall for it, even though you are so proud to have found a really nice item and would love to exercise your bragging rights.

After a close examination of your found jewelry, the spotter reports back to the "claimer" an exact description of the jewelry you've recovered and just shown to the spotter. A few minutes later, the claimer will then come up and say they were talking to someone near them down the beach or another part of the park and he (she) said you found a nice (ring / bracelet / watch / necklace) that may be theirs.

The "claimer" will then describe the item to you as if they had owned and worn it for years, and ask if you've found it. If you say no, then they act all indignant and start yelling that the other person saw you had what looks like the item they lost. They may threaten to call the police if you don't immediately give it back.

If you answer yes, that you have in fact found an item of that description, then expect that you will be asked to give it to them. The only way to defeat these people is to never show what you've found to anyone except a trusted friend or family member.

When people are close by and you find a nice jewelry item, it is always the best idea to keep it concealed in your scoop or palm the item so that it can not be seen until you can remove it unobserved.

If someone asks if you will help them find an item they've lost, ask for a detailed written description, as mentioned above, before you begin the search. If you find it, great, give it back to them. You'll feel very good about yourself and you will have done wonders for the hobby's reputation.

As a detectorist, expect folks to be interested in what you are doing and to stop and ask you about the hobby and / or the machine you are using. Always be polite. Try to answer as many questions as you can without taking up too much valuable time.

If you think the questions and answer session is going on too long, smile and tell the person you are hunting for an item lost by a friend and you have a limited amount of time to search, thus must be moving on. Folks will understand that and respect your efforts. If they ask what your friend lost, tell them prescription glasses or a car key, as most folks have no interest in beating you to a pair of useless glasses or key.

As previously mentioned, when detecting in public places like the beach or a park, realize that almost all children and many adults find what you are doing fascinating. Expect to be stopped to answer questions, the most common being, "Are you having any luck today?", "What is the best thing you've ever found?", and "What have you found today?".

You may enjoy hearing about a question asked one of my detecting friends and the response he gave one day at the beach. My buddy was working the moist sand at the water's edge and I was trailing behind him about 20 feet working in the water about knee deep.

I happened to look up and see an absolutely stunning young lady about "20 something" wearing a very skimpy bikini walk up behind my buddy and stop a few feet away to watch him as he was moving his detector from side to side to pinpoint a target. He scooped up the target and shook the sand from his scoop. As he was peering into the scoop, the gal walked up behind him and tapped him on the shoulder. He turned around and about broke his chin as it dropped to his chest when he saw the stunningly beautiful gal.

She asked him, "Does that machine find treasure?" He instantly responded, "Well young lady, I used to think so, but with you standing there, … it ought to be smoking right now and it isn't!".

She just looked at him and shook her head from side to side and told him in a semi-disgusted tone, "Cute, real cute", and walked away to continue walking down the beach. Oh well, I guess that response wasn't the best for advancing our hobby's reputation!

Improve Your Chances

If you want to hear about where jewelry may have been lost, cultivate the friendship of park and beach maintenance workers and lifeguards.

The best way to do that is to be obvious about properly disposing of the trash you recover when recovering targets or trash you see laying on the ground. Go to the lifeguard and ask where the nearest trash receptacle is located, if the location isn't too obvious. Show the lifeguard your pouch full of trash and perhaps a fish hook or piece of broken glass and mention how glad you were to have found it before someone stepped on it.

Seek out the beach or park maintenance person as they are picking up debris and again show them you are helping them do their job. Show your pouch full of trash stuff and ask if you can empty it in their trash bucket they are carrying. Try to strike up a conversation by telling them how on prior visits you saw some folks metal detecting, not filling in their holes and throwing trash back down they had dug.

Tell how angry that makes you and how just a few ignorant people can give the hobby a bad name. Suggest the worker not be shy about nicely asking folks to fill their holes and carry off the trash they find if they want to continue detecting on this beach or park.

You'd be surprised how often park and beach maintenance workers and lifeguards hear about lost items and how quickly they will tell you about them if you are perceived as "a good guy / gal".

Detecting Safety

Wear light weight work gloves. Work gloves protect hands from discoloring soil, broken glass, hypodermic needles, fire ants, yellow jackets, and the occasional baby snake you'll dig up.

If you treasure hunt in poisonous snake areas, snake chaps, snake boots, and keeping one of the headphones off the ear is a good idea.

Watch out for poison ivy, poison oak and any other stinging nettle type plants. Caution should also be taken against spiders around old home sites as well as in the woods. Slather on SPF 30 or higher sun screen on exposed skin to protect from eventual skin cancer.

Around abandoned old home / ranch / ghost town sites, there may be a water well covered over by wood or other debris, so be careful not to fall in. If a lot of old wood boards are laying about, watch for exposed nails.

Even if you are absolutely certain of surroundings, it is wise to use the buddy system when detecting. You never know when you may need each other's assistance for some medical or mechanical reason or to make you a lesser target to someone who intends you harm.

Also, it is much more fun to share the detecting experience and treasures found with a friend. Like airplane pilots and boaters, file a travel plan with family or friends as to where you'll be and when you can be expected to return.

When detecting in remote areas where there may be a possibility of being attacked by a bear or other dangerous critter, it may be wise to carry and be proficient in the use of a very large caliber pistol.

It is not wise to carry any form of concealed or visible firearm when detecting more urban settings where you may be confronted by persons who intend you harm. The buddy system is your best protection. A cell phone quickly brought out as if you are making a call is another good defensive action. Your second best protection would be your digging / recovery tool should you find yourself in an unavoidable situation where you are forced to defend yourself.

Before leaving on a detecting outing check that you have all required equipment in the vehicle ready to go along with spare batteries, a first aid kit, and something to eat and drink during breaks you should take on a regular basis. A freshly charged cell phone is nice to have in your equipment for a variety of reasons.

When detecting sites where explosive military ordnance may be found, such as abandoned WW1 and WW2 firing range impact areas (in Europe there may be unexploded land mines, grenades, and bombs) and sites of Civil War battles and gun emplacements, be extremely cautious when digging signals. Unexploded ordnance may still explode ***even if handled correctly by a trained professional***. Some fuses in shells and cannonballs are mercury triggered and a slight movement from digging or recovery may cause it to explode.

Do not attempt to defuse a cannonball or any other potentially explosive device by soaking it in water or oil. They have been known to explode even after weeks and months of such treatment. Black powder is particularly unstable after so many years and becomes liquid-like and gaseous ... very dangerous, very explosive, very deadly. It can explode destroying an entire home and everything in it!

If such an explosive device is found, leave it alone! Flag the site for future disposal by a professional from the police or the armed services of our country. Immediately report the location of the explosive device to the proper authority.

Should you desire to keep an unexploded cannonball or military shell you find for your collection, there are people licensed to handle and defuse such materials. For a fee, will come to the site, remove the ordinance and carry it safely back to their facility for disarming and stabilization against further decay. Obviously, this is an expensive service.

Asking Permission

Every piece of land is owned by someone or some organization. Never go on private property without permission. Always make sure you have permission before entering or detecting property other than your own.

If the landowner is not expected to be present when you actually plan to search the property, ask for a simple handwritten permission note in case a watchful neighbor or relative questions you being there.

Salt water beaches are generally considered public property and are open to detecting unless posted otherwise due to it being a National Monument, National Park, or National / State Historical Site. When in doubt ask local civic or park officials.

Churches are generally tax exempt corporations owned by the congregation or their governing body. They are not public property and permission must be gained prior to detecting there. Just because you are there ready to ask permission, and no one is there to give it to you, don't do it! Wait until a church authority is there and ask.

School yards and playgrounds of older schools have been productive for detecting. ***Do not go there during school hours***. Save detecting at schools for weekends and school breaks. Don't assume you have permission to detect there and argue your rights with the principle or maintenance person. You will lose! When appropriate, ask permission of that school's authority. You might even offer to put on a demonstration or class on artifacts, history, and metal detecting before you even ask for permission so they can get to know you better.

Receiving permission to detect on private property takes time and practice. Sometimes you will be rejected and other times you will be welcome to detect private property. Don't be afraid to ask permission. The worst thing that can happen is that the property owner will say no. If you never ask you'll never get permission. Here are some tips that have worked for me.

The best way to gain permission is to "network" for good contacts through friends, relatives, work, church, civic, and community organizations. The more people you meet, the more possible detecting contacts you'll make. Always keep a relic or artifact or two that you've recovered in your car or pocket to show anyone who voices any interest at all in your new hobby.

Sometimes when you've gained permission to search someone's yard in an old section of town, the neighbor may become interested in what you are doing and start asking questions. Take the time to introduce yourself and explain in a friendly matter of fact way what you are doing and why. Offer to help that person locate anything they may have lost in their yard or perhaps something the prior owners may have lost.

When "cold calling" for permission to detect an old home site you have been admiring, imagine how you'd feel if some stranger came up to your door asking to dig up your yard!

First, look presentable! Be wearing fresh clean slacks or jeans and a sport shirt. Park you car on the street or where it can be easily seen in the driveway. Leave your detecting equipment in the car. If you have a friend with you, only one should go to the door, the other person should remain in the car. When asking for permission, ***never*** have more than one other person with you.

Ring the bell or knock on the door, stepping away from the door to be as non-threatening as possible. When the door is answered, smile and speak slowly and distinctly introducing yourself and quickly explaining the purpose of knocking on their door:

"Hi, my name is ______ . I've admired this wonderful old home for years as I've driven by. Are you the owner?"

(Play it by ear at this point).

"Great, would you mind taking a few minutes to tell me what you know about the history of this property? I've really become interested in this area's history since becoming involved in metal detecting and researching the neat old stuff I've found?"

Again, play it by ear as they may have questions about you or the hobby. After you've heard what they have had to say and have had a chance to ask questions, thank them.

At this time, offer your time and experience with your metal detector to find anything they may have lost in the yard or flowerbeds. Ask them if they ever looked for old buttons or coins that may have been lost in their yard over the years. If not, would they mind terribly if you looked for a half hour or so with your metal detector. Offer them the chance to follow you around and see how your machine works and to show them anything you might find.

"Thank you so much. You've been very thoughtful taking your time to tell me about your home. If it would be of assistance to you, I'd like to offer my help with my metal detector to try locating anything you or a former owner may have lost while working in the gardens, flowerbeds, or yard. While we are at it we may find an interesting old button or coin. Do you want to try?"

Never use the word "dig" when talking about recovering targets. "Dig" has a negative connotation to the land owner meaning, "unsightly holes." If asked, explain that you *recover* targets by making a small slice to make a three sided plug in the turf and then pull the target out with the tip of your recovery tool. Offer to "water / fertilize" the plug after you fold it back into place.

When it's time to leave, thank the homeowner. If you've found more than one coin, offer one or more to them. Usually they will refuse the offer and thank you for your thoughtfulness. Take time to write down the land owners name and address to send them a thank you card.

You might also use this opportunity to ask if they have any friends or relatives who may also own property where you might search for old items that may have been lost over the years.

If your site to be searched is an occupied residential property, and if you gain the owners permission to search it, *never* have more than one other person with you on the first visit. Perhaps later, the owner may allow an additional person or two to join you in future searches.

If you work a larger site such as a farm or ranch that you'd like to return to time after time, it's a good idea to show your appreciation to the landowner by presenting a gift. A honey baked ham, or perhaps a bottle of nice wine (if you've seen evidence the landowner would not be offended by a gift of wine), or basket of fruit. At the very least, always send a thank you card.

Your thoughtfulness will be appreciated and may impact on your continued permission to search. It may also lead to you receiving permission to detect on other area sites.

Who owns items found on private property?
In most states the current landowner owns the rights to whatever you might find and is the only one who can give permission to keep what you find. If in doubt, check with a local legal authority.

You might ask yourself why bother detecting if you can't keep what you find? My answer to this is the fun is in the search and recovery. Like mountain climbing, you don't own the mountain, just the memory after the climb.

When we do metal detecting to help others recover items they have lost, we take photos of the finds and of giving back the lost item to the owner. More on this topic later.

When you are asked by the land owner to give them everything valuable you might find, you may want to suggest splitting the value 50 / 50. You likely will not run into this unless you are assisting someone to find and recover a hidden container of coins, jewelry or bullion someone had buried years ago. If this is the situation, a legal agreement should be drawn up by an attorney for you and the property owner specifying the percentage split and who maintains the "treasure" and where it will be secured until both halves are sold ... before the search begins.

It generally is not a good idea to stuff a legal agreement into a property owner's face when asking for or just receiving permission to search a new site.

Who owns items found on public property?
Items found on public beaches, parks, and school yards can be kept by the finder in most areas of the United States. If you find a class ring or some other item where it is possible to track down the person who lost it, make a good effort to locate the owner and return the item. You never know when that person you helped might be able to help you gain permission to detect on additional sites.

Some areas have laws that require anything found on public property over fifty years old to be reported to an authority. You should check to be sure of your rights as we encourage all treasure hunter's to obey all laws.

Detecting In Foreign Countries
Before metal detecting in foreign lands, it is advisable to find out what the regulations are for the area you plan to visit.

In some countries it is illegal to have a metal detector in your possession. In others you may have to purchase a permit and / or file export papers to bring allowable finds home. A good starting place to reference detecting laws of various countries would be the UK Detectornet web site, (http://www.ukdetectornet.co.uk/) This is one of the leading United Kingdom detecting sites. Click on their guide to laws in European and Asian countries. Before planning a trip to another country, it is advisable to check with the proper governing agency before booking the trip.

Detecting National Parks / Battlefields / Historical Monuments
Don't even think about it! Don't do it! Don't bother asking for permission as it will always be a gigantic waste of time to ask.

Never consider sneaking out and detecting these areas without permission. When caught (not ***if*** caught), you will face a fine of many tens of thousands of dollars, many years of jail time, and confiscation of all vehicles and equipment, and all relics you have ever recovered that may be on display or stored in your home or workplace.

Once, when detecting with permission, on one of only four remaining pieces of private property within Gettysburg National Park Battlefield, our three person group was visited by two Park Rangers. We were asked if we had permission to detect on the private property. We showed the Rangers our written permission from the owner since the owner was not home during our visit.

The property owner had cut the grass to his property lines and we had placed marking flags two feet inside the property lines just to be certain we were not on battlefield property. We were thrilled to be detecting within site of Little Round Top, Big Round Top, and the Peach Orchard.

With a big grin on their faces, the Park Rangers told us if they "caught us on the National Battlefield property we would do more hard time than if we hacked up our neighbor with a butcher knife!" They even recorded our names and took photos of us as well as photos of the car we were using. We were asked if we would be on other private property in the area over the next few days.

Once assured we were "good guys", the Rangers told us of the hundreds of people they catch and prosecute each year who pleaded ignorance of the law. They told us how they have folks in civilian clothing patrolling the park on foot and in unmarked cars searching for people trying to steal the National Battlefield treasures still under the ground. They also patrol at night in all kinds of weather using heat imaging and night vision equipment to catch illegal "nighthawks."

If you are interested in helping to recover relics from government owned historical sites, realizing you will never get to keep even one item you find, contact the Park for the name of the person who is responsible for Park archeology. If you have a metal detector in your car when traveling to or through a National Park, I urge you to keep it disassembled and the batteries removed.

Ask if you might volunteer time and equipment under their supervision the next time they schedule a project requiring soil to be turned or moved. Almost always, the parks use students of archeology from the major universities serving the area and you will not be allowed to participate. Smaller, less noted, more remote Parks may welcome your assistance under their close supervision. If you ask, the worst they can say is no thank you.

Loaning, Gifting, or Selling

Most treasure hunters find themselves accumulating coins, jewelry, and artifacts they find during the practice of the hobby. At some point in time it makes sense to empty out the boxes of stuff you've recovered and put your items on display for others to enjoy.

If you don't have adequate space in your home or workplace,

consider loaning or gifting a display of items to your local museum or library.

If you decide to loan a display of artifacts, make a careful written and photo inventory of what is in the display and have an agreement that your name be displayed as the owner and a method established so you can be contacted by interested persons to have questions answered.

Should you decide to give a display to a museum, specify under what conditions it is to be gifted. You may consider an agreement that specifies the return of the inventory to you should the display not be kept in an area easily accessible for public viewing or perhaps sold to another museum.

Selling artifacts has become big business since the advent of Internet auctions and retail antique stores. People have received thousands of dollars from selling war relics from the Revolutionary War era through modern wars. Valued collectables can be almost anything from horse shoes, political buttons, clothing buttons, pocket knives, coins, jewelry, children's toys, give away premiums from food products, tokens, writing instruments, old weapons, hat pins, watch fobs, etc.

There are two schools of thought on selling artifacts found while detecting. Some of us are against it, and some of us are for it.

Those that are against it say it hurts our hobby and eventually will lead to fewer rights to find and recover items from history. Once an item is sold and placed into a private collection, the public is deprived of viewing it.

A number of historians and archaeologists would rather see a relic rot away in the ground from never being recovered rather than have "an amateur" find it and sell it or keep it themselves for non-public display. This is a good reason to discourage selling and to give items on a regular basis to a local museum. A little cooperation may go a long way toward improving our relationship with historians and archaeologists.

There are places available to make reproductions of more valuable finds. Oftentimes a museum can tell you whom to contact for this service and may even pay for the reproduction if you give the original to them.

You certainly wouldn't want a very valuable item on display in your home or workplace where it may be stolen. A good reproduction can be viewed and enjoyed without danger of it being stolen if it is identified as a reproduction.

Those that sell items think that the money they make helps offset their equipment, research, travel, and time expense incurred during practice of the hobby. To some, this turns the hobby into a business and eventually becomes work. I certainly wouldn't want to quit my day job in hopes of making a living by metal detecting. Only you can decide what is right for you.

Treasure Hunting Code of Ethics

Western and Eastern Treasures Magazine publishes the following code of ethics each month and encourages the ethics be strictly followed as well as being in strict compliance with all local, state, and federal regulations and restrictions.

I will respect private property and do no treasure hunting without the owner's permission.
I will fill all excavations.
I will appreciate and protect our heritage of natural resources, wildlife, and private property.
I will use thoughtfulness, consideration, and courtesy at all times.
I will build fires only in designated or safe places and only if needed for safety and or security.
I will leave gates as found.
I will remove and dispose of any trash that I find.
I will not litter.
I will not destroy property, buildings, or what is left of ghost towns and deserted structures.
I will not tamper with signs, structural facilities, or equipment.

Metal Detecting Organizations

In many areas of the country there are organizations for the purpose of promoting the hobby of metal detecting and preserving the rights of detectorists. If you would like information on finding the most local organization and metal detecting events they may offer, contact the Federation of Metal Detector & Archeological Clubs:

FMDAC
117 Forest
South Hutchinson, KS 67505
(313) 665-8693
www.fmdac.com

Our Valued Reputation

"It only takes one Ah Shucks to wipe out a thousand Atta Boys."

I'm sure you've heard that old adage before. This is true as it relates to the hobby of metal detecting. All detectorists are encouraged to be aware that our actions, both positive and negative, reflect on everyone else involved in the hobby.

When people see us on the beaches, parks, or fields practicing our hobby, they will have one of two thoughts if they have never experienced metal detecting themselves. The first thought can be one of fascination. "Gee, that person is treasure hunting. I've always wanted to do that, I wonder what he or she has found? That looks like so much fun!"

Unfortunately, the other thought can be, "There goes another scavenger digging up our pristine beaches and parks taking advantage of someone's misfortune. They ought to outlaw those people".

Sometimes in our excitement or anticipation of finding treasured items, we may unknowingly do something, we didn't realize at that time, may reflect badly on the hobby. Just as there are good and bad drivers on the roads, there are good and not so good treasure hunters practicing our hobby all over the world. Hopefully the following thoughts will help enhance your reputation.

From time to time newspapers and television news feature human interest stories about metal detecting. The majority of these features shed a very positive light on our hobby. Others do not. The best stories are usually about a treasure hunter helping someone recover an item that had been given up as lost forever. Often the lost item has more sentimental than monetary value.

We have all read about a class ring being found and of the treasure hunter's efforts to track down the high school or college from which it was purchased then to track down the person who lost it via initials on the ring. Usually a few local or long distance phone calls later, the owner or relatives of the owner can be reunited with the lost item. Nothing was charged for the favor.

I remember an article from a paper that was meant to be positive but in fact reflected badly on the hobby. The article had a photo of a gentleman detecting on a beach and another photo of the accumulated rings, bracelets, necklaces, and coins he had found over the years. I can't remember exactly, but the headline was equivalent to "Treasure Hunting For Fun and Profit."

The main point received from reading the article was if you ever lose something valuable in the area this gentleman works, contact him for help. He would ***only*** charge you $45 per hour plus 15% of the items value if found.

You can imagine the problems this "pay for help" idea may cause between the detectorist and the person who has asked for help. A disgruntled customer who thinks they may have been taken advantage might even report this "business" to state and federal licensing / taxing authorities. We all know what level of fun, paperwork, and taxes that can bring!

There is nothing wrong with keeping valuable items if you are unable to locate the owner. Gosh, we have a display here in our store of jewelry people have lost that we were not able to trace the owner.

Perhaps I'm wrong in thinking our hobbies reputation is not enhanced by turning it into a business where a person's loss can be made even greater by paying an hourly fee should the item not be found. I realize we have expensive equipment in which we have invested, batteries to buy, vehicle and gasoline expense, and even tolls to pay in many areas of the country.

You do incur expenses in trying to help someone through our hobby. Most of us have no problem at all in asking that our actual out of pocket expenses incurred helping someone be reimbursed.

You also have expenses associated with playing golf, league sports, fishing, water skiing, bicycling, or any other recreational activity yet most folks wouldn't even think about turning those hobbies into a business.

Those who detect ***primarily*** for profit justify their position by thinking it is only fair for them to offset all their expenses plus make a profit "when they help someone." I can understand their position, I just do not agree with it.

All of us give back to our community via church donations or donations of time and / or money to many worthy charities because it is the right thing to do. Later in this book we discuss a program you might consider for your area that allows detectorists to help people while at the same time discouraging nuisance requests for your time in finding relatively worthless, easily replaced items. Look for the section titled, *Do The Right Thing*.

Proper detecting etiquette that enhances our hobby includes honoring the "space" of others who may also be treasure hunting or involved in some other activity where you may be searching. This is particularly important in parks or on beaches where folks have personal possessions laying about that could easily be stolen.

You would not appreciate someone coming right up to your beach towel or picnic blanket to metal detect regardless of you being close by or not. Should something be stolen from another person's personal possessions, and you were seen detecting within their space, who do you think might be accused of the theft?

Respect the space of another detectorist that is working an area you would also like to detect. Stay far enough away from another treasure hunter so that your detectors do not cause interference with each other. The units may experience chattering and false signals if operating on similar frequencies.

Use headphones when detecting to keep target signals or the machine's audio threshold tone from bothering the peace and quiet of others. Headphone use also results in less battery drain than relying solely on the built in speaker.

Smile and say hello (or at least nod) to others as you pass by on your search. Take time to talk with people asking about the hobby and what you may be finding. Believe me, they are intrigued with what you are doing or they wouldn't be asking.

When asked what I'm finding, I smile and tell people "I'm finding aluminum pull tabs from cans, a few coins, and when lucky, a piece of jewelry." Since I use a White's detector with target ID, I delight in showing people how it can identify a quarter or other coins and rings with the Icon picture and VDI number. I also like to show them how the machine tells depth of the item too.

One of the first questions asked is usually, "What is the best item you've ever found?" Once you have had a chance to find some expensive jewelry or perhaps an interesting old artifact, this is a fun question to answer.

A big benefit from being friendly to the people who ask about your new hobby is that you'll sometimes make a new friend who may invite you to detect on their property. I average about ten detecting leads a year just by being nice to people when they have asked about the hobby. Some of the leads resulted in worthwhile items being found, others did not.

We can't stress enough how important it is to fill in all holes you dig and removing all trash you find and disposing of it in a refuse container. Most of us will wear a two pocket nail apron or a soft bag that has pockets for trash and treasure. Smokers will want to snuff out their cigarette butts and put them into the trash pocket rather than throwing them on the ground.

Many of us become very frustrated when we observe others, who when detecting, do not take the time to repair any damage from recovering a target or don't wear a nail apron or pouch to hold the trash they recover until it can be put in a trash can.

When I see this happening, I approach the individual and introduce myself and attempt to strike up a short conversation about the hobby and finds being made.

At the appropriate moment, I mention that I saw them throwing their trash targets back down on the ground (or not filling their holes) and that is a good way to get all of us banned from detecting that land. I suggest they use their pockets until they can come to the store for a free nail apron and give them my card if I don't have a spare apron in my car to give them at that time. That way, the person knows I'm not just being critical, but want to be helpful. Most folks won't get angry. Some folks just aren't the "sharpest knife in the drawer" and will tell you to mind your own business.

Either way, you've made your point and the person usually starts filling holes and saving the trash for proper disposal.

Do The Right Thing

Because many of our customers know I enjoy the hobby of treasure hunting, and because we sell metal detectors here in our store, we are often contacted by people who have lost items they would like found. They ask directly for my help or if they can rent a metal detector and try to find the item themselves. We didn't want to charge people for our service as we thought it would be bad for our hobby's reputation.

When time was available, I would help people for no charge at all. When I couldn't go search because of business obligations, I would contact one of my detecting friends to help … also at no charge.

We looked upon assisting these people as a challenge and a test of our skill with our equipment. If time was of the essence and no one was available to help, we would rent the person a detector.

We really do not encourage renting detectors because it is so time intensive to train someone in its use. We also don't want to turn a potential future detectorist off on the hobby because they failed in finding the item they were after.

As our reputation spread from a number of articles in the newspaper and TV features about us helping folks, we were starting to get a lot of calls for help to find easily replaceable items such as car keys, blazer and band uniform buttons, or small tools and parts. Some of the items we were finding were not worth enough to justify renting a metal detector.

We were using up a lot of quality detecting time and burning a lot of gas in the car. We knew we had to do something to stem the growing number of nuisance requests. Our unofficial small group of treasure hunters got together and agreed on the following ideas.

First we agreed that it should be worth a donation to a local charity of at least $25 to someone in the city and $50 far outside the city for our time and efforts regardless of if we found the lost item or not. Our minimum cost to rent a detector at that time was $40. We agreed we would accept personal and business checks and no cash for our efforts.

We further agreed that we would ask the person to make out their check payable to the Cape Coral Police Department's Do The Right Thing program. The program is led by our Chief of Police to reward area youth for making "right" decisions in their life, especially when it sometimes may not be the most popular decision among their peers. The program receives little city funding and relies heavily on corporate sponsor and individual donations.

We contacted the Chief's office for permission to perform our voluntary metal detecting services for check donations to the Do The Right Thing program.

The Chief was delighted with the idea and helps us promote the service.

Norm Jones returns a ring to the owner.

Upon our request, he wrote us an introductory letter on his letterhead to give to property owners asking them to allow us to park on their property while searching for items lost on or near their commercial or private property.

We have found most people donate more than the minimum we ask. On occasion, after handing us the donation check, they almost fight you trying to stuff a $20 or $50 bill in our pocket for our efforts when we have found their treasured item. What the heck, if they are that happy and want to force a modest cash reward on us, we'll accept the money to offset some expenses.

Some of the people we have assisted have helped us gain permission to detect on property that was unavailable to us before. When we detect in parks around our town, most police officers now smile and wave and drive right on by instead of slowing down to ask what we are doing. The whole concept is a win / win situation for the hobby, our area youth, for us, and the people we help.

As you gain more metal detecting experience and skill, friends and associates will remember you because of your hobby and give your name to others seeking help in finding their lost items. We encourage you to start a similar program among your group of detecting friends to do good in your own community.

Hopefully, you too will enjoy the gushing tears of appreciation and a big hug in returning a well loved and sentimental item to the rightful owner.

Through our program we have found expensive his and hers wedding and engagement rings "tossed in the heat of battle" and lost over four years before we were contacted to find them.

We have recovered keepsake rings that have been handed down for generations that had been lost for several years.

We have recovered a man's wedding ring that was lost during his honeymoon three days after being married and quite a few engagement rings over the past few years.

$15,000 ring returned to the owner.

A 50 + year old woman received our help trying to prove to her doctors that there was a bullet lodged in her head that she said was fired into her by her father when she was a child. It hadn't shown on doctors X-rays and CAT scan nor did an XLT Spectrum metal detector reveal any presence of metal other than tooth fillings and ear rings! We suspect their may have been absolutely nothing worth finding in this woman's head.

We were able to determine that a cute little three year old girl had in fact swallowed a penny and that the parents would be looking at a couple more days of "treasure hunting" until the coin was passed. Because of our help, the parents would not have to expose their daughter to unnecessary X-rays.

Don't worry about taking a check made out to the police department. Should it bounce, they will be dealing with the police department, not you! Check with your local police authority to see if they have a similar program in place you can support. If not, pick out the local Rotary or Kiwanis Club to support or perhaps the local food bank for the poor.

Cleaning Coins and Artifacts

Modern coins are quite easy to clean using a rock or coin tumbler with water, cleaning powder and aquarium aggregate. Coin shops sell special cleaning products that may be more expensive than necessary but do a really great job. The idea is to clean the coins up enough to be spent in stores, parking meters, or vending machines.

Do not clean silver with copper coins. Clean them separately as the silver coins will take on a copper sheen if cleaned with copper coins.

The value of old coinage and artifacts will be dramatically reduced if polished or cleaned incorrectly. Use pure virgin olive oil to soak old coins and artifacts to slowly dissolve the accumulated dirt and grime that has built up over the years.

Figure on soaking most items in pure virgin olive oil using a lidded plastic container for at least two weeks before attempting to scrub them with a medium to hard nylon bristle tooth brush. For heavily encrusted items you may need to repeat this process several times over a year or longer.

Realize some items become so corroded over the years from synthetic farm / lawn chemicals and natural galvanic corrosion that images on them may never become fully visible.

Never use acids or electrolysis on anything except ferrous metal or modern coinage. These cleaning methods ruin the value of brass or other non-ferrous relics.

Don't Throw It Away Until You Are Sure

I almost learned the hard way to not throw anything away until you are sure the recovered item is trash. Let me give you an example or two.

Some time ago I was visiting a friend in England so that we could go metal detecting. One day I recovered an item that first appeared to be some type of brass or perhaps bronze metal harness part from a horse drawn wagon or farm equipment.

It was oval shaped with a rectangular piece attached that was perpendicular to the oval. I threw it beside our car thinking I would later throw it in my trash sack in the trunk of the car as we left the field.

Later I picked it up to examine it again as I had thought it may be a forearm retainer from a musket. Upon closer examination I decided it couldn't be from a musket since the rectangular piece that I thought would have contained the musket sling, was not made to swivel and was turned the wrong way for a sling to be attached.

The oval shape was also too small to go around a barrel and wood forearm. I then placed it in my trash side of my apron thinking it was probably some harness part or some other form of trash.

On the same field I found a curious looking rod shaped piece of bronze with a ball shape on the end. The ball shape had several curved decorative yet primitive grooves on it. None of the three of us that were detecting that day had a clue as to what the item was. We just knew that it was not new. It may have been old trash so I put it in the trash side of my apron.

When I got home from the trip I discovered both items were still in my apron so I soaked them in olive oil to clean them up as best I could. I decided to take photos of both items and send them off to someone who might be able to identify them. I was later surprised to learn that both items were not trash but now two of the most favorite items I keep in our store artifact display.

The oval shaped item was identified as a Medieval purse ring. The rectangular part was where it was worn on a belt. Purses at that time were drawstring pouches. The top of the pouch was closed with the drawstrings then pushed up through the loop and secured by tying off the drawstrings to the loop. Pockets in clothing were not common until the 16th Century.

The second item was identified as "the main portion of an ornamental bronze pin in use somewhere between 100 BC to 100 AD."

Another item found that same day was a tiny, almost oval shaped flat piece of bronze with a bit of pinched metal protruding from the back. After throwing it away, I decided to keep it just in case it might be worthwhile. I later saw a line art drawing of the same item in a British metal detecting magazine identifying it as a Tudor period clothing pin.

Here's a Y2K disaster! A metal detecting friend has concentrated on coin hunting since 1970. Like most "coin shooters" he also keeps other interesting looking finds. Those finds are eventually cleaned and tossed into a detecting odds & ends box. In 1972 he was searching the yard of an old residence and found a round object with a funny looking eagle on it. Not knowing what it was, he chunked it into the box.

About 1982, under pressure from his wife during their annual Spring cleaning, the junk boxes were sorted and most of the stuff thrown away. In the year 2000, he met another metal detectorist whose passion was finding Civil War relics.

In looking over his new friends collection, lo and behold, there was a round thing with a funny looking eagle on it proudly on display. His friend had identified it as a Union Army breastplate. He suddenly remembered he had also tossed a round thing with a funny looking eagle on it into his detecting junk box.

Not remembering what had been kept and what had been tossed during the Spring cleaning over 10 years ago, he hurriedly returned home to search his junk boxes for what he now knew was a Union Army breastplate ... remembered to be in beautiful condition ... and worth several hundred dollars. He searched through all his detecting junk boxes several times to no avail! The only tiny bright spot in this story is than he did keep the Union Army Cavalry button found in the same yard.

If an item is old, keep it until you are sure it is not trash. Even if you wouldn't want to display the item, someday you might find someone who would want it for their collection.

Competition Hunts / Club Hunts / Rallies
Well organized active local / regional metal detecting clubs may sponsor club hunts or competition hunts. In some parts of the world they are also called a rally. When new to the hobby, joining a club is a good way to develop new detecting friendships that may become regular detecting partners.

An active, well organized club may require each member, or a committee of several members to research, find, and secure permission to detect at least one site each year upon which the entire club may search as a group. The site may or may not be available again in the future for a club hunt or for individual detecting. Depending on the specialty of the club, if any, a club hunt site may be a potential relic site, old coin site, coin / jewelry beach site, or military / historical site.

Some club hunts are limited to members of that club only, while other hunts may be open to members of other clubs as well. Some club hunts are designed to invite new detectorists to become a member of the club. Club hunts are a good way to get to know the other club members better and to pick up detecting / recovery techniques from those that are not as new to the hobby.

Some club hunts may require a participation fee be paid to the organizing club. The fees collected may be paid in total to the land owner, paid to the landowner's favorite charity, split between the club and landowner, a portion of the fees may be paid out as prize money, or kept in total by the sponsoring club. Items recovered most often may be kept by the finder.

Some club hunts on historical sites may require the targets be "flagged" as a potential relic and left in the ground for later recovery that day or the next day by the finder, under supervision of the governing authority. Items found are kept by the governing authority for study and eventual public display.

Competition hunts will generally have prize tokens or "bought" coins and jewelry planted to be found by the hunt participants. Often, cash / merchandise prizes are given to the finder of specific prize tokens or number of tokens / coins found within a timed period. Competition hunts generally require an entry fee for different levels of participation. Some competition hunts or events are team based, others are individual. Some competition hunts offer fun ways to get to know others in the form of game hunts. The following are examples of some fun competition hunts.

<u>Poker Hunt:</u> Tokens are made up for each card in a deck of cards. Each participant pays an event entry fee plus an "ante". The ante becomes the prize for the winning 5 card or 7 card hand. The idea is to find the tokens which make up the winning hand of poker. During

the event a finder will yell out the token "card" they just found and hope to remember who is yelling out what card so at the end of the event, you can find the other participants that hold the cards required to win and split the event's cash prize.

Other versions of this event may require a team 7-card stud approach where at timed intervals, the team gets together to determine if they have good enough cards to want to make a bet, "see" another teams bet, or even raise the prior bet and continue in the event … or "fold", letting the remaining teams continue to build the best possible 7 card hand.

Traffic Light Hunt: Painted lead sinkers or tokens of red, amber, and green are planted in the ground all over the event field. The first person to recover all three colors is the winner. The challenge here is knowing that there are far fewer of one of the colors planted than the other two colors. It can be made even more challenging by having each of the three colors painted on differing types of metals.

Three Legged Event: Two detectorists share the use of one metal detector and recovery tool. Naturally, they have their legs bound together. A clear path is marked from the start to finish line and each path has the same number and type of targets buried in it. To win, the team must learn to effectively and quickly swap off using the detector and recovery tool to cover the entire width of the path until the required number of targets are recovered.

When participating in this event, make sure that the partners machine that isn't being used is clearly marked with the owners name. It can be a nightmare to determine who owns what machine after the event when so many of the same brand and type are being claimed from the storage area when the event is finished.

Target BINGO: Four to six pairs of matching numbered tokens are buried along with lots of other numbered tokens that have no match. The object is to find a numbered token and then match it with its same numbered token found elsewhere on the field by another detectorist. As tokens are found, pandemonium starts after about five minutes with participants shouting numbers all over the field. The one to six people that find all the numbered pairs equally share the cash prize for that event.

<u>Treasure Chest Hunt:</u> A locked treasure chest is displayed by the event field. Inside the chest is the winners prize. Keys are scattered all over the event field. Sometimes all keys are on the surface, sometimes keys are shoved just under the grass. Only one or two of the keys will open the treasure chest depending on the value of the prize. When a key is found, the detectorist runs to the chest to try to open the lock. The more valuable the prize, the more people will want to enter the event with a chance to win.

Club hunts are usually smaller with fewer problems with the participant's detectors causing interference problems with each other. Competition events, if well organized, with a good prior reputation, and well advertised may draw hundreds of participants. These type's of competition hunts may not be as enjoyable for those that don't enjoy crowds.

Before signing up for an hunt, ask all about what you may expect as to the hunts total cost, individual event costs, and location. Ask about rain date plans, expected number of participants, what type of hunt it will be (researched relic site / newly planted coins and tokens), events, will overnight accommodations (motel / camping) be available, can meals / snacks be purchased, will there be evening entertainment.

If it's a newly planted old coin / tokens hunt, ask if nickels will be included so you know how low to go on discrimination. Ask what metal the tokens are manufactured from. Will targets all be shallow? Will all coins / tokens be buried flat or will some be on edge?

Before leaving for the hunt, make sure your detector has a fresh set of batteries and is in excellent operating condition. If you can arrange with someone to have a back up detector to use if yours should fail, that's a good idea ... if you are familiar enough with the back up machine to use it effectively. Double check against your WRITTEN equipment list that you have packed everything you'll need.

There are two schools of thought on hunt coil sizes. A larger coil can cover more ground but is more difficult to quickly pinpoint close together targets. A smaller coil covers less ground but is quicker in pinpointing targets. A larger coil finds stuff deeper and covers more shallow surface area.

Practice setting your machines discrimination to ignore everything below the lowest phase target you expect to recover if only shallow targets are planted. Some depth loss can be expected from higher discrimination.

Practice quickly finding and recovering a lot of targets close together at home by scattering coins in an area. Include some shallow buried coins so you can practice with the tool you'll use in the event. In most hunts where you are searching for newly planted coins and tokens, the majority of items are found and recovered within the first half hour or less so it's very important to learn to work quickly.

If you have a friend with a detector that is of the same frequency, or has a multiple frequency detector, have them crank it up within a few feet of you so you can practice listening to your machine sounding off on known targets along with the interference you'll be listening to from the other machines operating near you.

If you don't do this you will become very frustrated during the hunt and lose time and targets looking for a place to hunt with no interference. Don't look for such a place as it just won't be within the hunt's boundary. Expect to hunt with interference and practice dealing with it before the hunt!

Ok, you are now at the hunt and have attended the pre-hunt informational meeting and know the boundary's and rules and are waiting for the starting signal. Crank up your detector, ground balance / set discrimination, adjust headphone volume, etc as you want to be ready to detect as soon as the signal sounds.

The hunt begins! Keep yourself and your detector in motion. When you get a signal, recover it and throw it in your pouch without looking at it. You'll have time to admire each coin / token later. If possible, keep your coil in motion while bending up and down to and from recovering a target as your next target may be just a few inches or feet away.

Keep your eyes focused on the ground, not on what others are finding. Pay no attention to a different recovery technique than the one you are using. It's too late to begin an unpracticed new style of hunting during a competition. When the number of targets recovered begins to slow down, you may want to glance around to see where groupings of people seem to be concentrated and most seem to be recovering targets. That may be a hot spot of concentrated targets you might want to explore if it's not too far away.

Ignore folks who want to talk to you and concentrate only on swinging your coil and digging your own targets. There will be time to talk with other participants later. Remember to dig every signal within the depth parameters of the event that are above your acceptable discrimination level. Yes, you'll recover some trash, just throw it in the bag and keep moving and detecting. If your machine has a depth reading, that information will be more important to you than the targets ID, so don't waste time deciding to dig on ID, ... just focus on depth decisions and pinpointing.

Display of Coins, Relics, and Artifacts

We encourage all treasure hunters to display your better finds where people visiting your home or workplace may see what you've found and enjoy it as much as you. Craft shops or picture frame shops will usually have a selection of table top or wall mount display cases. If you are handy with tools they are not difficult or expensive to make.

Here in our store we have converted a large antique Case® Knife display and a couple tool bit display cases to display what we have found. We are now running out of display space and must buy or build a larger metal detecting finds display cabinet.

It is helpful to viewers to type up a brief identification / explanation of the less obvious items. If you have a digital camera and computer you can take photos to show the reverse side of rare coins or other interesting pieces where needed.

How To Identify Old Coins and Relics

Look in your local library for specialty books on the items of your primary detecting interest. The major book sellers in your area as well as on the Internet all have quite a few reference books available.

Visit local museums to see what types of old items they may have identified and on display. Talk with the museum's director to ask for help in identifying the more interesting finds you make.

Ask to speak with someone in the local university's history department for help in identification or as to where reference materials may be found in their library. You may also find they are interested in learning how to metal detect and may have a source of good sites to search as well.

Best Advice For New Treasure Hunters

Experienced detectorists were asked what advice they would give to someone new to the hobby. We think their responses are interesting.

"There will be good days and bad days. Don't become discouraged on the bad days. Always remember, a bad day detecting is better than a good day at work."

"Read everything you can about the hobby."

"Don't become impatient, approach the hobby one step at a time. Study it inside and out and enjoy it. The good stuff will come."

"If you think you are going to become rich with this hobby, find a different hobby. If you like the excitement of finding something from the past regardless of it being a rare coin, relic, or just old junk, and you have the patience, then this is the greatest hobby there is with the ***CHANCE*** of finding something very valuable for your collection or even for sale."

"The rewards are not always the targets found, but the people you meet, the exercise you get and the sites you see while detecting. It is hard to establish a value on these benefits of the hobby."

"Have a long talk with someone in the hobby for years. At first your local dealer can steer you in the right direction. You can learn in a few minutes what it took an experienced person weeks or years to learn. Why reinvent the wheel?"

"Talk with as many old timers in your area as possible about what they remember of their past or their parents and grandparents past, and the community (s) in which they have lived."

"Read as much as you can about the history of your area. The local newspapers have a wealth of information in old copies of their publications. There is usually a historical society in most cities and towns. Become involved and as you develop research and metal detecting skills and accumulate interesting finds, make presentations to as many groups as possible. Through presentations to schools, colleges, Rotary and Kiwanis Clubs, you'll meet people that can provide you with future detecting leads and sites."

"Spend the time necessary, perhaps 100 hours or more, to learn about what your metal detector is telling you."

"Before recovering the target take the time to sweep the loop again and hear what the machine sounds like. Remember target identification and audio information the machine is providing. Then recover the target. Oftentimes the target will give a slightly different reading out of the hole and laying on the ground versus un-dug laying in the ground."

Starting with a cheap "no name" unit can result in quickly becoming discouraged. They are almost worth what you pay. Today's detectors are so much more sophisticated than the ones used from the 1960's to today's sophisticated machines. A machine with target identification, either digital or analog, along with depth readings can help you be more productive quicker.

Spend time on Internet metal detecting forums. You can gain a wealth of information from forums. As with any public forum you will have to consider the source of the information provided and judge for yourself if the information is accurate and worthwhile or just from someone with an axe to grind or some other motive in providing the thoughts shared. Overall, your time on the forums will be worthwhile. The primary forum reference for beginners should be the one sponsored by the manufacturer of the detector you bought. Why spend time learning about how to more effectively use brand A when you need to learn about brand B.

When talking to people who are advising you what brand to buy, every detector brand out there "is the best detector ever made" depending on what the personal preferences and experience has been of the person giving you the advice. There is a learning curve involved in whatever brand you start with and the brand you may add to your detector selection in the future.

There is no one detector that "does it all the best" under all conditions. If you enjoy the hobby as much as most of us, you are very likely to buy at least one more detector. After you buy it, don't be surprised if you don't like it as much as your first detector until you learn all the benefits and limitations the machine has to offer.

Buy a machine for its proven performance in the type of hunting you want to do. Learn to use and learn to like every machine you buy.

Digital Photography and Detecting

At some point in time you may be lucky enough to receive permission to detect on relatively virgin territory that may yield items of historical interest or value. So valuable and rare they should be on public display in a museum. Perhaps you received permission to detect at that historical location based on your promise to donate all that you find to the property's managing organization.

Wouldn't it be nice to record your finds and the detecting experience with photos? Digital photography allows you to review photos right after they are taken to insure photo quality is acceptable. Photography allows the best of both worlds. You gain the detecting experience of a lifetime and have pictures of the experience to enjoy forever.

Photos are almost as much fun for people to see as the real thing when making presentations about the hobby and what you've found. Use photos to record your collection for insurance purposes should a loss be experienced through catastrophe or theft.

Those of you who live "up north" where it gets too cold to go detecting for months at a time may find manipulation of digital photos you've taken during treasure hunting expeditions an interesting diversion until your season warms up enough to start detecting again.

There are Internet forums just for the purpose of sharing photos of items found by treasure hunters.

Bragging rights are a fun part of the hobby among detecting associates as well as your non detecting friends. Sometimes the proof of your skills will be from the photos you take.

The cover and internal photos of this book were taken with a digital camera then sized and enhanced with a photo editing program.

Finding the "Big One"
Let me share an true story example of realizing the dream shared by treasure hunters everywhere. Harold and Pearl Kelly were once owners of a local pub in their small town in England. Harold started to develop signs of heart problems and was advised by doctors to retire and take it easy.

They now had time on their hands and decided to involve themselves with metal detecting for its exercise and fresh air. The Kellys managed to go out for a few hours several days a week and began to find artifacts and coins of all types from iron age to modern.

I had the pleasure of treasure hunting with Harold and Pearl during trips to the UK. They would get me excited and encouraged with reports of treasure troves and hoards of silver and gold coins that had been found by others. They always said that one day, they too hoped to find "the big one".

As I left to go back home to Florida that day in late July, I wished good luck to them, thanked them for sharing their best sites with me, and told them not to forget to send me photos of the big hoard when they find it.

In mid December of that same year I received photos of Harold and Pearl with 150 silver and 1 Roman gold coins from the period of Emperor Julian, 360—363AD. They were found by the Kelly's on one of the same fields, in the exact same part of the field we detected together in July. They had gone back the first week in December after the farmer plowed the field the last week of November.

Over the next few months an additional 39 coins were recovered. The find is referred to as "The Wentbridge Hoard" and is on display at the Museum in Doncaster, Yorkshire, UK.

The photo to the right is a close up of a plow damaged hammered silver coin thought to be from the same hoard. It was "hammered" (minted) in Lyons, France. It was found about a year later by Harold in the same general area. Note the image of Emperor Julian on the front.

On the reverse side is the word VOTIS followed by the Roman numeral for 30. Under that is the word MULTIS followed by the Roman numeral for 40.

VOTIS was the number of promises made to the Army when Julian was trying to become Emperor. Promises such as shorter marching days, better food, higher pay, warmer uniforms, better weapons, etc.

MULTIS was the number of filibuster length speeches Julian supposedly made to the Roman Senate trying to keep the promises made to the Army.

If the Senate thought the Emperor was spending too much on the Army, they had him killed. If the Army thought the Emperor was treating them unfairly, they killed him. This was the main reason Roman Emperors held office an average of less than two years.

Julian and other Emperors of this period changed the VOTIS and MULTIS numbers on their coinage to suit their own political needs of the time. As savvy politicians, the coins in use in Rome oftentimes had lower VOTIS and MULTIS numbers than those issued to the Army thousands of miles away in another part of the world. Some things just never change with politicians!

"Must Have" Accessories

The detectorist that has ***everything*** even remotely necessary will be able to fill the trunk of the car and a closet with all the equipment. Many accessories are required for different types of metal detecting under varying weather and terrain conditions.

It is truly an awesome spectacle to watch an avid detectorist arrive at the site and prepare for the hunt. It could easily be compared to a Roman Gladiator, knight in armor, or the DELTA FORCE preparing for battle to the death.

Clothing is often changed, added, or removed based on conditions. Equipment is selected, strapped on, put in pouches, bags, pockets, back packs, or hung on belts. Is all this stuff really necessary? Probably not, but until you gain your own comfort level as to minimal equipment needs, we offer the following suggestions.

Extra batteries are essential to have in your equipment package if you will be detecting more than 10 minutes away from where replacements can be purchased.

Quality recovery tool (s) speed up and make target recovery easier with little to no damage to the target. For land detecting, a sharp "knife style" trowel with one shape edge and a root cutter on the other edge is highly recommended. In farm fields, a short "D" handle shovel or folding entrenching tool will be the tool of choice.

Yes, you can use grandma's old rusty garden trowel, but when you finally test your buddy's "official metal detecting recovery tool", you'll end up putting grandma's old rusty trowel back in the garage where it belongs. Hey, we aren't talking planting pansies. We are preparing to more easily recover valuable objects without damage in tough soil conditions using the least amount of time.

If you will be beach or water detecting, a heavy duty metal hand held scoop or long handle scoop will save you much time and effort. If you will be prospecting for nuggets, a pick (with magnet attached), shovel, and small tough plastic scoop will make the day more productive and pleasant.

In softer, less rocky soil a ***brass rod coin probe*** may be handy at times to "feel" for that old silver or gold coin in the bottom or side of the recovery hole. This isn't a "must have", but more of a nice to have accessory.

The item to never be without when detecting away from the beach, is an ***electronic pin pointer***. Popular brand name pin pointers include Bullseye, Merlin, and Vibra Probe. Each of these pin pointers are hand held and emit a tone and vibration as the probe's tip nears the target hidden in the plug you cut, the dirt pile, or still in the recovery hole. Each of these three have different types of on / off switches, length, weight, and detecting distance from the probe's tip.

Unless you've used a hand held electronic pin pointer, you really cannot appreciate how much time and effort (energy) a pin pointer will save you. I'd drive an extra hour to return home and then go back to the site if I forget to take along my pin pointer for a full days detecting!

Modern metal detectors will respond on metal targets so tiny they may be as small as #6 birdshot from a shotgun shell. Small targets can be so encrusted with dry or wet soil as to not be felt or seen without the aid of an electronic pin pointer to reveal their location. Trust me on this one folks ... this should be second only in importance to a quality recovery tool!

Some type of ***pouch, bag, or nail apron*** to hold trash and treasure is mandatory. Specialty pouches and bags made just for metal detecting will also hold your pin pointer, digging towel, probe, cell phone, and even a water bottle.

Knee pads worn when detecting protect you from nasty bruises from stones and staining grass or soil. Knee pads help keep you dry when the ground may be wet. I personally prefer softer sponge rubber pads over the type of pads with a hard plastic outer liner. Velcro straps on the pads make them quick and easy to put on and take off.

A ***smaller coil*** than the standard size coil is a must for detecting in high trash areas. Smaller coils help the metal detector isolate on the good target in the ground near a trash target.

A ***coil cover*** makes a lot of sense to protect your coil from damaging nicks and gouges when detecting frequently in rocky terrain.

Headphones / Extra headphones being available have saved the day for many detectorists. An extra set of headphones is secondary in importance to having one good quality set in excellent working condition. Good headphones allow you to hear some deeper, initially "iffy" signals that you won't hear with a metal detectors built in speaker. I refer to the built in speaker as the "Ah Shucks" speaker because it should only be used when headphones are not available.

Headphones use less power than the detectors built in speaker making your batteries last longer. When selecting headphones, make sure they fit comfortably so that your ears or head won't get sore. Select headphones that have a "curly" cord. Straight cords tangle too easily. Do not use a plug adaptor to convert a mini-plug to 1/4". Purchase headphones with the correct size plug for your machine.

A "nice to have" but not necessarily a must have accessory is a ***hand held GPS*** (Global Positioning Satellite) device. This is the same type of device that tells operators of modern boats, cars, and airplanes their exact location anywhere they may happen to be.

Why is this important? Someday you will be out driving around scouting for new sites, or while detecting see a place you'd like to return for another hunt. A GPS can instantly record the latitude and longitude of the location usually within less than 50 feet. A GPS can provide your current location should you ever need to call for assistance.

Mistakes That Make You Look Stupid Or Cost Money

I'll make one exception about not revealing stupid personal mistakes. Here is one that my friends remind me of every time we detect a park.

Tuesday is my day off and the day I try to go detecting each week. It was early in February and the temperature here in SW Florida was a chilly 49 degrees as my buddy Larry and I arrived at a city park maintained soccer field right as the sun was coming up. I was wearing jeans, a jacket, and gloves to keep warm until the sun would do this for me.

After walking about half way across the soccer field, I received a head banging signal with target identification numbers and depth indicating it may be a gold item. The week before I had found a nice gold bracelet with almost the same target ID. I carefully got down on my haunches to avoid getting my knees wet from the dew covered grass. With the digging towel spread out to contain any soil particles from the target recovery, the digging tool was then carefully pushed through the turf to begin the target recovery.

Suddenly, over top of the headphone's threshold tone, I hear this huge "SWOOSH" and was almost instantly bowled over by a huge stream of very cold water from the parks underground pop-up irrigation system.

It became obvious this system was designed to deliver a gazillion gallons of water a minute with penetrating TURBO power to a distance of at least one hundred yards. The sprinkler head was seemingly laser targeted with smart bomb accuracy right up my backside from less than two feet away!

Sheepishly, I looked up to see Larry laughing so hard he had tears streaming down his otherwise warm dry cheeks. Trying to regain lost dignity in a hopeless situation, I suggested to Larry we move to the other soccer field until the sprinkling cycle on this field was complete.

As we were walking to the other field (well Larry was walking, I was squishing along), I pointed out to Larry how I had used this somewhat rare opportunity to wash off my digging towel, digging tool, and my detectors coil (without getting the electronics wet) at the expense of the city parks & recreation department. The rest of the morning was spent with me facing west to insure the warm morning sun was on my soaked backside.

I'm not going to tell you which of the following tips result from my personal long list of things I try very hard not to repeat. Trust me, some of these are based on experiences of mine, others were told to me by customers and some I witnessed and chuckle about from detecting friends.

Electric fences make entry into some detecting fields quite inconvenient, sometimes painful, and always shocking. One must also not assume a rusty fence is OK to use as a target or to sit upon when answering nature's call.

If a target can be detected under a fully functioning electric fence by your metal detector without falsing and chattering, do not lean forward while recovering the target and accidentally touch your forehead or hands to the wire. Be extra cautious of this on hot sweaty summer days.

Do not lean your metal detector against an electric fence.

Do not assume all those signals you are receiving when detecting under a high tension / high voltage electric transmission line or near a TV station's regional transmission antenna are actually targets to be recovered.

Do not become so excited about a coveted recovery you just made deep into the thick old growth woods ...and that you want to show to your friends right away ... that you forget where you laid your detector down before seeking your friend (s) perhaps a couple acres away.

Never assume a docile looking cow, munching on grass, acting like it is not interested in you, is actually not interested in you. Never turn your back to cattle when recovering a target.

And since we are on the topic of cattle, metal detecting, and related notes of interest, never assume a "cow pie" that looks completely dry on top, is dry all the way through and not very slippery. Never assume there are no good targets to be dug out from under "cow pies." Sometimes there are!

If you lose the nylon bolt or nut that secures your loop to the down rod, go to your nearest hardware store and buy a nylon toilet seat bolt. Do not use any type of metal bolt or nut as you will, at best, make it impossible to pinpoint a target using the detectors "all metal" pinpointing mode. At the very worst all you'll get from your detector is constant screeching.

Never assume your metal detector is "falsing" or broken when every hand full of soil you pass in front of the loop when recovering a target from the hole causes a good signal. Try taking off all jewelry from that hand and wrist and then see how much better your detector works.

In your excitement to start out on your treasure hunting trip or in your haste to leave the field at the end of a good long day, don't assume your metal detector is packed away in the vehicle where it should be. If you feel a slight bump and hear the sound of crunching bending metal as you back up, that probably means you forgot and left your detector leaning against the back bumper.

If you reach your detecting spot or arrive back home without your metal detector, that probably means you missed seeing it in your rear-view mirror as it slid off the roof of your car and started to bounce and slide down the road. For what its worth, a sticker with your name and phone number on it might be good to place on your detector.

Never confuse the terms "weatherproof" with "waterproof" or assume they mean the same. They don't!

When searching shallow water without using a waterproof metal detector, don't hold the wet dripping loop over your head while walking from one location to another. Water will run down the down rod sometimes entering the electronics.

Never assume a shallow looking "dirty water" pond or stream has gently sloping sides to give you plenty of shallow underwater bank to detect. It is a good idea to bring along a spare set of dry clothing.

It is much easier to pinpoint a target near your feet when not wearing steel toe safety shoes or boots.

Never go metal detecting without a extra set of fresh batteries nor forget to turn your detector off at the end of each trip.

In your excitement to return to treasure hunting after a coffee or lunch break, don't forget to turn your metal detector back on before continuing to search. With your machine turned off, I guarantee you won't find anything and will discover it is a huge waste of time and quite embarrassing to cover the same area twice.

Don't assume your metal detector is broken because it does not turn on and you just put in fresh batteries. Check the polarity positioning of each battery to make sure the +'s and -'s are ALL pointing in the correct directions.

Never assume your metal detector is faulty because all it wants to do is screech at you when you turn it on. Check to see if your pinpoint switch is in the "always on / all metal" position. If it is, click it off and the screech will go away.

When planning to travel to detect in a foreign country, take along a set of electric plug format converters for your re-charger and a step down 220 to 110 volt converter. The little popping sound you hear and the slight wisp of electrical smoke you smell can be immediately traced to the 110 volt battery charger plugged into a 220 volt outlet using the plug format converter ***...without using the step down power transformer.***

When re-charging at the end of the day, in your thirsty anticipation of a couple pints of "wallop" at the local pub followed by a great pub cooked dinner … don't forget to plug in the power converter to the end of your plug converter ...that's plugged onto the end of your charger ...before plugging everything into the 220 volt outlet.

Do not attempt to flail a snake to death with your metal detector. You will most likely break your detector and / or get bitten in the attempt. Yes, snakes are very scary to most of us and we don't like being surprised by one under a bush or rock around which we are detecting.

I have been told by snake "experts" how nice snakes are when given a chance ...and that a snake is as afraid of you as you are of it. I really doubt the information snake experts give on this is correct or if snakes actually listen to snake experts!

Do not attempt to pin a snake to the ground with your detectors loop and then shoot it with rat shot or a slug from your pistol. Again, you will likely damage your machine and turn what (according to the experts) was a nice scared snake into a very irritated snake with intent to kill you and then eat you.

When detecting in an area known for being rattlesnake habitat, keep one ear free of the headphones. If you have hearing difficulty, keep your best ear uncovered! Wear snake boots and snake chaps to protect from sneaky snake attacks.

No matter how good it sounds, ***never*** dig a signal from under an active fire ant hill. Do not kneel down on your knees when recovering targets from areas where there are fire ant hills. Do not wear shorts and shoes with no socks in the same type of area.

When detecting in an area known for having a bear population, always detect with a partner. ALWAYS choose at least one detecting partner that runs more slowly than you.

Never assume the horse watching you detect from the other side of the fence likes to have his nose scratched. On that same line of thought, never assume the farmer's dog will remember you have permission to be there from your last trip. If that same dog reluctantly allows you out of the car to detect, and then later invites himself to share your "brown bag" lunch, do not argue with him!

Leave your detector on the ground when needing to quickly climb a tree. Remove your headphones before climbing.

When hunting in salt water, never assume that what bumps into you is probably friendly, not hungry, or at the very least not angry you are in it's way! Always wear some type of foot protection.

When water hunting in southern states, don't bet you can wade faster than a snake or alligator swims. You can't, so remain observant!

When planning to detect a public park or athletic field that has an in ground sprinkling / irrigation system, it's good to find out the time and days the sprinkler heads are scheduled to operate. Don't assume you can run faster and further than a fields "TURBO" powered pop up irrigation system can deliver water. Along this same line of thought, don't assume high pressure water delivered from an industrial sized sprinkling system does not hurt when blasting you from a couple feet away!

Top 10 Clues You Are Too Involved In Metal Detecting

Within a relatively short period of time, you will realize the hobby has become so intriguing, that you seem to always be practicing the hobby, thinking about, or planning for it. A metal detecting friend from an Internet detecting forum sent me his list of the top 10 clues that you may have become too involved with the hobby. Thanks to David Colburn, Emmett, ID for this list:

#10 The detecting equipment required by you and your detecting partner won't fit into a standard sized vehicle.

#9 All of your trousers including your best Sunday suit have grass stains on the knees.

#8 The floor and trunk of your vehicle are always littered with beer can pull tabs and bottle caps, and you don't even drink.

#7 You wear out the toes of your shoes before the heels or soles.

#6 You find that when you take a family outing to a fair or carnival, you take along your GPS to record the coordinates of the wild rides and ticket booths.

#5 The front and back door in your home have full jars or buckets of modern dirty coins used as a door stops.

#4 When walking, you always catch yourself looking down.

#3 When you hear your Minister tell the congregation, "Please be seated", you respond with, "Amen Brother, or at least a Barber!"

#2 You think discrimination is a bad thing mostly because it keeps you from finding small gold.

#1 A friend proudly tells you his new girlfriend has a nice set of double "D's" and you naturally assume he means she purchased 2 more special coils for her metal detector.

Airlines / Cruise Ships and Metal Detectors
Terrorist attacks around the world have resulted in heightened airport / airline security. When planning to take your metal detector on an airplane or cruise ship, it's best to disassemble your machine and pack it in your luggage. Pad the machine well using clothing.

When checking your luggage in at the airport, tell the counter agent that you have a metal detector in your luggage in case they may want to inspect it right there at the counter or perhaps just mark the bag with a special tag.

You can also carry your machine on board as a "hand carry" if you are willing to show your machine is in fact an operating detector at EVERY security checkpoint. Remember, almost all airlines restrict passengers to one carry-on that will fit into the overhead luggage rack or under the seat in front of you.

NEVER EVER attempt to bring your digging tools, "pig sticker type" brass rod coin probe, or any other equipment that could even remotely be construed as a sharp or pointed weapon on board as a hand carry. Pack that type of equipment in your checked baggage.

Airline security has dealt with traveling detectorists for many years and they are familiar with the equipment. I've traveled domestically and internationally via commercial aviation with my detecting equipment since the tragic attacks of 9/11. Yes, the security checkpoints are much more interested now than ever before. Yes, they will ask you to insert the battery pack and show them the machine works and is not filled with explosives. Yes, the extra security is a good thing!

If you have any doubts at all, contact the airline / cruise line and ask in a positive manner, ***"What is best method to assure my metal detector will arrive at my destination undamaged the same time I arrive … as checked luggage, or as a hand carry-on?"*** Make sure you write down the name of the person giving you the information.

Your metal detector is an electronic device no more sinister than a laptop computer, cell phone, or digital camcorder and should be treated as such. If in the very unlikely event you are deigned transport of your equipment by a counter agent or security checkpoint employee, in a friendly manner, ask to speak with the supervisor for a second opinion. Ask the supervisor why your equipment may present a danger to fellow passengers when laptop computers and cell phones do not.

Trivial Information: Gold Markings

Ron Smith of Alpine, TX is an avid treasure hunter. One day he found a gold ring that wasn't marked with the standard karat rating commonly used in the U.S.A. The gold ring carried a marking of .417.

Ron did some research and found that some European countries use a rating indicating the percentage of actual gold present in the alloy used to make the ring. He found that .417 is the same as a 10k marking. Here is the table of values Ron found in his research:

.333 = 8k
.417 = 10k
.583 = 14k
.666 = 16k
.750 = 18k
.917 = 22k
1.00 = 24k (pure gold)

Trivial Information: Squirt Now, Play Later

How frustrating is this? You spend hours upon hours of time doing research to develop leads that may eventually point you to a detecting site with great potential. You do further research and find the owner of the land. After days and days of playing phone and door tag you receive permission to detect the property.

Finally you are able to coordinate the day and time of the hunt on the great new site with your detecting partner. You leave well before dawn for the hour or longer drive to the site to take advantage of the cool crisp morning and the remaining hours of detecting time before dusk.

The site may be deep into the woods or perhaps just hidden from the road by dense vegetation. Regardless, you park the vehicle, gather all the equipment needed and start walking to the site.

After a relatively short walk, you immediately see evidence the site in fact may have super potential for finding old coins or relics. Perhaps a cellar hole or foundation stones can be seen, the old chimney is still partially intact, or perennial flowers are in bloom where an old flower garden used to be located. What's even better, there are no mounds of modern day metal trash laying about. That's the good news!

The bad news is that the site is so overgrown with briars, bushes, brambles and waist high grass, it may be impossible to search effectively, safely, or completely. The good news is you have an alternative spot selected to detect and even better news is that you noticed a hardware store just a mile or so down the road from this site. It is now time to prepare for "Squirt Now / Play Later".

Go to the hardware store and purchase a couple quarts or gallons of RTU (ready to use) grass & weed killer such as Triox® (total vegetation killer for up to 1 year) or the more common Round Up® brand of grass and weed killer.

You travel back to the site and spray the most promising areas to kill all the vegetation. The site may be in good enough condition to make your initial hunt within a week or two. If you kill the weeds at the end of the growing season, you'll have a longer time to hunt the site over the cool late fall and early spring months.

If you are planning to hunt the site again during growing season, take along additional vegetation killer to further prepare the site for future trips if need be. Also remember to take along a machete or long handle grass & weed whip to help cut dead material from where you may want to hunt.

Remember, this is not your land so kill no more vegetation than necessary and place your cuttings into a neat pile out of the way.

Trivial Information: Treasure From The Sky, Meteorites
It is estimated that over 50,000 meteorites hit the earth every year. Meteorites, depending on their type can sell from a few dollars per gram to thousands per gram. Meteorites are quite popular as a collectable. When meteorites sell for that kind of money, it places them into a value category that can exceed gold and platinum.

Some metal detectorists specialize in searching for these high iron / nickel content chunks of metal from outer space. Meteorite chasers research reports of impacts and some will travel the world to find them.

One such chaser, Ed Milota of Search & Recover Enterprises, located in Ventura, CA told me he uses a White's XLT Spectrum programmed to accept VDI numbers of –74 down to –94.

In November of 2001, Ed sent me a meteorite he found that weighs 16.2 grams and is about one inch by three quarters of an inch in size. It looks like a dark brown and beige stone. Had I found it, I would have thrown it away thinking it was a "hot rock". You can bet that from now on I'll save all stones / rocks / pebbles that I find that react to my metal detector until it is determined the object is a meteorite or not.

To have any chance at all of finding and identifying a meteorite, you need to learn all you can about them. Learn what the various types look like as to colors, shapes, sizes, weight, metal content, and how deep they can be expected to be found in the area you most commonly hunt. There are quite a few books published about the subject. Read as many of them as you can and carry a color illustrated small pocket edition with you for help in identification.

If you have a near-by museum of natural history, spend time looking at their display. Even better, talk to the museum's curator to tell of your interest in finding meteorites and that with the museums help you will be more successful and will donate a sample or portion of the larger meteorites you may find. Perhaps you can arrange an appointment with the curator to let you air test their various samples with your detector to learn how it will respond at specific settings.

If that is not possible, ask if the curator knows of any local collectors who may be able to help you to learn how your machine might signal on a meteorite.

Wouldn't it be a shame if you detect in an area known for its meteorite impacts and didn't know why your machine was responding to an occasional rock or stone?

Trivial Information: Diamonds In The Sun

The insurance industry estimates over 100,000 diamonds are lost from jewelry settings every year according to "Jim" from an Internet forum I visit from time to time. Sorry Jim, I don't know your last name, but remembered most of what you had posted about your winter time diamond hunting when you weren't able to go metal detecting.

When Jim is reluctantly persuaded to go to the mall so his wife can shop, Jim spends his time looking for lost diamonds. He wears polarized sun glasses and says diamonds sparkle in the sunlight while glass appears dull.

He searches the grout joints in the tile at the mall entrances where people may lose a stone while removing gloves. He also checks around the drains in parking lots at the mall and at nightclubs. He says the mall janitor told him of the technique. Jim found a half karat diamond in this manner.

Other folks will cruise mall parking lots in winter looking for lost jewelry as the snow melts from each storm. Particular attention is paid to the areas where the huge piles are melting from where the snow plows pushed it out of the way.

Something Old, Something New

If you've ever married, had daughters that married, or are thinking about getting married, eventually you will hear the following bride's preparation rhyme. It is a checklist of what the bride needs to be wearing during her wedding ceremony to help insure the marriage is long and happy.

"Something old, something new,
Something borrowed, something blue."

I was reminded of this bride's rhyme in England after finding a worn Six Pence silver coin, bent with one edge up, the opposite edge down. I thought it was just a damaged coin. I later learned it is a "love token". They are not rare but certainly not uncommon finds in England. Occasionally one will be found here in the U.S.A. in areas once influenced by the early British settlers.

When a young man became enamored with a young girl and thought she had mutual feelings, he would give her a "love token" as a symbol of his love.

Today, we call that "going steady", or becoming engaged. The girl would keep the love token so long as her love for the young man continued.

Six Pence bent into a love token.
Charles III, 1696, United Kingdom

When the relationship "broke up" the girl would throw away the love token. When the relationship resulted in marriage, the love token would be worn in the bride's shoe during the ceremony. It was a good luck omen for a long and happy marriage. It was the bride's first evidence of her young man's unending love. Thus, the last line of the bride's rhyme is actually …

"A six pence for my shoe."

The United States Of America began minting coinage following the Revolutionary War. British coinage and most other things so blatantly British quickly fell from favor right after the war. Six pence coins were no longer in circulation in the U.S.A thus the supply eventually disappeared. The last line of the bride's rhyme was dropped over time as well.

"Something old, something new" is certainly what our wonderful hobby is all about. Every item of value or interest we recover has a history. That history makes our finds unique and the hobby so very interesting.

Finding Civil War mini balls should make one wonder how terrified a soldier must have been to fire that round in battle in an attempt to kill someone … or to have one fired back. Dropped rounds, some say, have no history. I disagree. A dropped round may have been caused by running and reloading in terror in a hail of gunfire. A dropped round may have been the result of a young lonely soldier playing with the mini ball in camp out of boredom. Perhaps it was dropped when the bugle sounded the call for another terrifying attack.

Old coins, old jewelry, buttons, watch fobs, medals, and tokens, all have their own story as to how they ended up under the coil of a metal detector.

A sparkling diamond engagement ring or wedding band found at the beach may show wear from generations of mothers who wore it and handed it down to their daughters. A tearful young lady in great emotional distress can certainly be imagined when a new engagement ring is recovered. A base metal child's ring, pendent, or toy car is no less of a loss to them at their young age.

Those of us that have been involved in the hobby for a number of years and find the hobby more fascinating with each trip. You will too as you learn to take maximum advantage of all the features you metal detector has to offer. Over time you will develop greater research and detecting skills and begin finding some really interesting objects.

You are urged to read everything you can on metal detecting, study the owners manual to learn your detector inside and out. Spend as much time as you can detecting with an experienced treasure hunter. Do research for the purpose of locating sites that may be lesser known or perhaps even unknown to experienced treasure hunters.

Share your good sites with an experienced person and learn all that they might be able to teach you in exchange for you taking them to your better sites. Have an understanding that these good sites are to be kept exclusively for use only by you and the person (s) you invite. This is a normal and acceptable agreement within the hobby.

When invited by another treasure hunter to detect on one or more of their productive sites, respect them by not revealing those sites to others or by not detecting there without his or her permission.

Do not become discouraged by others who may have fancier machines than you and who seem to find more treasure and less trash. They were new to the hobby at one time and they also recovered their fair share of trash targets.

Make a personal commitment to become the best detectorist and ambassador of the hobby that you can be.

Encourage, teach, and develop friendships with others new to the hobby as you become more proficient yourself. Every time you get to go treasure hunting, remember … "It's out there, all you have to do is go find it."

Metal Detecting Glossary

A proper glossary of terms should be alphabetic, making it easier for the reader to reference an unknown term to learn its definition. This glossary isn't! Hey, this is my book and I'll organize the glossary my own way so that related terms are somewhat together.

To all my middle school, high school, and college English teachers, some of which, through no choice of mine, I enjoyed more than one year, … most of whom that by now are almost certainly eternally resting , … please forgive me. I have once again broken your rules!

… "**Nah-Nee-Nah-Nah-Nah-Nah**!"

Hammered: Primarily refers to coins that were made by striking an impression onto a flan (thin disc) of metal by hand using a hammer and a die (i.e.: hammered silver, hammered bronze, hammered copper). This method was used to make coins before the advent of machinery produced coins in the late 1500's. Secondarily, used to describe the condition of someone who has consumed far too much adult beverage.

Relic: A memento or monument of the past. Not to be confused with treasure hunters that are retired with gray hair and who metal detect very slowly because that is the only speed left working in the bodies old gearbox.

Artifact: A product of human workmanship, especially of simple primitive workmanship. Often considered by treasure hunters to be interchangeable with the word "relic".

Barber: An old silver U.S. coin featuring the bust of some guy who looks like he needs a haircut.

Loop: The little round "thingy" at the end of the metal detector that detectorists sweep back and forth over the ground. Also referred to as a coil.

Coil: The little round "thingy" at the end of the metal detector that detectorists sweep back and forth over the ground. Also referred to as a loop.

Down Rod: The interconnecting tube (s) between the electronics control box and the loop / coil. Also called the rod, lower rod, or clevis rod among other nomenclature coined by the various manufacturers. Many down rods are sectional using cam locks to fasten the sections together. The section closest to the loop / coil is non metallic, made from a fiberglass or composite material.

Discrimination: Not related at all to social interaction or being politically correct. Discrimination in treasure hunting is a setting of the metal detector's circuitry to exclude non desired metal targets. Some metal detectors exclude all metals below the setting and respond with a signal on metals above the discrimination setting. More sophisticated metal detectors offer "notch" discrimination allowing the user to accept or reject specific types of metals across the entire spectrum of ferrous and non ferrous metals.

Sensitivity: The measure and control of a metal detectors ability to perceive changes in the different types of metal within the loops / coils detecting pattern. Can also be the measure of a treasure hunters ability to determine the level of anger the spouse may be demonstrating when the treasure hunter returns home from a treasure hunting trip … on the spouse's birthday … and you obviously forgot a birthday card, present, and the always important "Happy Birthday To You" song. The treasure hunter is considered very sensitive (and somewhat glib) if he or she can overcome the spouse's anger.

Hoard: A hidden supply. Treasure hunters seek a hoard of money, jewelry, and or precious metals.

Cache: A hiding place. To put, to hide to store in a cache. Early explorers would store implements or provisions in a cache.

VLF: **V**ery **L**ow **F**requency. A metal detector operating in the 3 to 30 kHz frequency range. Mid range is considered by many to be about 6 to 10 kHz. Dual and multiple frequency metal detectors can still be VLF machines.

Ferrous / Non Ferrous: Ferrous is the classification of metals that are generally attracted to magnets such as iron and steel. Non ferrous metals do not attract magnets and are typically found by treasure hunters as gold, white gold, copper, silver, brass, lead, bronze, nickel, and aluminum.

LCD: **L**iquid **C**rystal **D**isplay. Used by many metal detector manufacturers to create a screen providing information as to the metal detectors set up and target / target recovery information. Tiny pixels form the graphics of numbers, lettering, icons, and / or graphs to present the information.

Analog: Refers to a metered scale generally using a needle to point to target and target recovery information on the meter for the user. Analog controls are dial knobs, push buttons, and switches.

Digital: Refers to the processing of information received from the coil. Digital controls are manipulated usually through a touch pad membrane scrolling through menu windows rather than through a mechanical control using analog knobs and switches.

Pinpointing: The process of locating a target under the center of the metal detectors coil or within an inch or so away from an electronic probe's tip prior to recovering the target.

Conductivity: The measure of a metal target's ability to allow eddy currents on its surface. Combined with "Phase", tells the metal detector to respond to a target or not and identify it depending on the discrimination setting and target identification features. If you are really tingling with excitement about this electrical stuff, see "phase".

Phase: The length of time between eddy current generation sustained on a metals surface and resulting secondary electromagnetic field effect on the loops / coils receiving winding. Combined with "conductivity", tells the metal detector to respond to a target or not and identify it depending on the discrimination setting and target identification features. If you know about and really understand phase and conductivity, you are either an electronic pencil necked techno-geek or are reading a book about detecting that is far too elementary for your level of detecting knowledge and skill.

Whatzit: Refers to a recovered target that the treasure hunter has no clue as to what the item may be. A whatzit becomes trash when disposed of properly. When disposed of improperly, a whatzit becomes litter.

Dunno: The answer to your metal detecting partners question as to the identity of the "whatzit" you've recovered.

Hot Rock: Unlike a hot car or hot guy / gal, a hot rock is useless, worthless, and a waste of time for detectorists. A hot rock contains a much higher concentration of mineralization than the surrounding ground. Hot rocks will more often than not cause your metal detector to respond as if it was signaling on a desired target. A good example of a hot rock is the byproduct "coke" from the coal fired steel manufacturing process.

Leaveitright: An obvious piece of worthless debris too heavy or large to carry away for proper disposal. "***Leave it right where you found it***." Not to be confused with a whatzit or dunno because you are absolutely certain the leaveitright has no use to you and is too heavy to carry away for proper disposal.

Clad: Refers to USA dimes, quarters, and half dollars made starting in 1965. Prior to 1965 these coins were silver. Since 1965 these coins are stamped from sheets with a core of copper or zinc and "clad" with copper / nickel / silver.

Pull tab: The finest example of the worst invention ever created by man. The pull tab comes in various sizes and shapes as it has evolved since the early 1960's from the original larger size with "beaver tail" to the modern smaller size riveted to the aluminum beverage container. Bored stupid careless people keep playing with the pull tab until it comes off the container then throw it on the ground to be later found and disposed of properly by a treasure hunter.

A direct derivative of the hated pull tab is the pull lid as exampled by Vienna Sausage and potato chip can lids. Unfortunately, pull tabs, nickels, and many gold rings can not be differentiated from each other by most metal detectors. You have to dig them if you hope to find gold … those that dig the most trash recover the most treasure.

Treasure: An understanding wife or husband that acts as if they are interested and share in the excitement of what you've recovered during a day of detecting as you explain in minute detail what the objects may be. Treasure is most often defined by non-detectorists as money, bullion, or jewels hoarded up in reserve for future use.

The very best treasures you will find are the new friends you'll meet and make through metal detecting! As far as the objects you will eventually find and recover, the most "treasured" are often the least valuable but the most interesting as to their history and use.

ICON: The digital or printed image on a metal detector's digital or analog meter that suggests the metal detector signals only on coins, jewelry, and relics. Upon gaining experience in the hobby, ICONS are enjoyed by the user for entertainment value only. They should be considered only as "eye candy". When demonstrating to others how smart your metal detector is using known good targets and pieces of trash, ICONS are very impressive. "Oh my GOSH! How does it know that?"

Air Test: An exercise demonstrating how far away a metal detector will signal upon an object suspended in the air. Unless one is metal detecting in weightlessness, all metal objects will be laying on the ground or buried under the surface.

Ground Test: A more realistic exercise in determining how far away a metal detector will signal upon an item buried under the surface considering the actual ground conditions you will be hunting.

HALO: For those of you who are into sky diving, this means High Altitude (departure from a perfectly good airplane), Low Opening of your parachute. A really reckless way to arrive more quickly at a destination. In metal detecting, HALO is also referred to as Aura.

Aura: In metal detecting, a metal object laying in the ground for a very long time allows the soil surrounding the item to take on similar characteristics of the object due to natural galvanic corrosion. The result of an item that has developed an aura or halo effect, is that a metal detector may signal on the object at a depth it would not signal on were the item newly buried.

Hot Rock: A highly mineralized stone that causes a metal detector to respond as if it were a desirable target. Coke, a byproduct of burned coal is a very common hot rock. Some fields are littered with hot rocks to the point where it is impossible to swing the coil without detecting one.